T0310443

Introduction to Industrial Polypropylene

Scrivener Publishing
100 Cummings Center, Suite 41J
Beverly, MA 01915-6106

Publishers at Scrivener
Martin Scrivener (martin@scrivenerpublishing.com)
Phillip Carmical (pcarmical@scrivenerpublishing.com)

Introduction to Industrial Polypropylene

Properties, Catalysts, Processes

Dennis B. Malpass
and Elliot I. Band

Scrivener

WILEY

Co-published by John Wiley & Sons, Inc. Hoboken, New Jersey, and Scrivener Publishing LLC, Salem, Massachusetts.
Published simultaneously in Canada.

For general information on our other products and services or for technical support, please contact our Customer Care Department within the United States at (800) 762-2974, outside the United States at (317) 572-3993 or fax (317) 572-4002.

Wiley also publishes its books in a variety of electronic formats. Some content that appears in print may not be available in electronic formats. For more information about Wiley products, visit our web site at www.wiley.com.

For more information about Scrivener products please visit www.scrivenerpublishing.com.

Cover design by Russell Richardson with concept and figures supplied by Elliot Band

Library of Congress Cataloging-in-Publication Data:

ISBN 978-1-118-06276-0

Printed in the United States of America

10 9 8 7 6 5 4 3 2 1

This book is dedicated to the memory of Professor Jerome F. Eastham, Sr., who lived his life with courage and élan. He was a mentor, a scholar and a true Tennessee gentleman. The world is a poorer place without him.

Jerome F Eastham, Sr was born September 22, 1924 in Daytona Beach, FL. He grew up in Lake City, FL and completed high school in Hazard, KY. He served in the US infantry in World War II. After the war, he attended The University of Kentucky and graduated in 1948. He then attended graduate school at The University of California in Berkeley and received a PhD in organic chemistry. Following fellowships at the University of London and the University of Wisconsin, he accepted an assistant professorship in 1953 at The University of Tennessee in Knoxville. During his early years at UT, he became well known for his work in organic and organometallic chemistry. He was a dynamic instructor and inspired many graduate students over his long tenure at UT. In the late 1960s, he made a bold decision to change careers and entered medical school to pursue an interest in neurochemistry. An author of this text (DBM) was privileged to be Prof Eastham's final graduate student in organometallic chemistry before Prof Eastham entered medical school. After obtaining his doctor of medicine in Memphis, Eastham returned to Knoxville where he became an attending physician in the Department of Internal Medicine at the UT Research Hospital. Synchronously (one of his favorite lecture words), he resumed teaching organic chemistry to new generations of students. He completed a career that spanned 39 years with the University of Tennessee and retired in 1995. He passed away on August 22, 2008.

Contents

List of Tables

List of Figures

Preface

Crystalline polypropylene was discovered in the early 1950s and commercial production began in 1957 in Italy, Germany and the USA. Since that modest beginning, polypropylene has become among the most important synthetic polymers produced by humankind, ranking second only to polyethylene. (Actually, polypropylene ranks first if polyethylene is separated into its various types, e.g., HDPE, LLDPE, LDPE, EVA, etc.) Estimates indicate that approximately 55 million metric tons (~121 billion pounds) of polypropylene were manufactured globally in 2011. Within the few minutes it takes to read this preface, about 600,000 pounds of PP will have been manufactured at facilities scattered around the world. Though production has slowed in recent years (especially 2009) because of the global recession, polypropylene has resumed its historic healthy growth. Polypropylene is manufactured in various forms on 6 continents and its applications are ubiquitous in daily life, from the fiber in your carpets and the upholstery in your living room furniture to the casings for the power tools in your garage.

The intent of this book is to provide chemists, engineers and students an introduction to the essentials of industrial polypropylene—what it is, how it is made and fabricated, how it is characterized, the markets it serves, and its environmental fate. Technical aspects are described in a straightforward way with minimal discussion of esoteric theory such that a person with a modicum of training in chemistry should be able to grasp. Our purpose is to provide practical, down-to-earth discussions of polypropylene technology. Extensive theoretical discussions are considered outside the scope of this book and have been largely avoided, but details are available in excellent handbooks and encyclopedia articles, many of which are included as references at the end of each chapter. Because of the industrial focus of the book, we also cite a number of relevant US patents.

Another key objective of this text is to supply perspective on recent innovations in the polypropylene industry. There has been tremendous hoopla about single site catalysts since their industrial use began in the early 1990s and the role they may play in revolutionizing the polyolefins business. That may well occur in the future, but, for now, these catalysts must be factually described as minor contributors to industrial polypropylene. Separating hyperbole from reality, we learn that single site catalysts are estimated to account for less than 3% of today's global production of polypropylene. Of far greater immediate import has been the emergence of innovative cascade processes (also called tandem or "hybrid" processes) such as Spheripol, Spherizone and Borstar (see Chapter 8). Nevertheless, SSC are technologically important and in Chapter 6 we shall address key features of single site catalysts and the cocatalysts commonly employed, including a recent innovation that renders the historic SSC cocatalysts unnecessary. We ask only that the reader keep in mind the relatively minor contribution of SSC-derived polypropylene to today's marketplace.

We also intend the text to be useful as a supplement to college courses on polymer chemistry. In addition, Chapters 2, 9, and 10 provide a practical basis for developing a laboratory program, which is a unique feature among texts on this subject. This book will answer fundamental questions such as:

- What are the principal types of polypropylene and how do they differ?
- What catalysts are used to produce polypropylene and how do they function?
- What is the role of cocatalysts and how have they evolved over the years?
- How are Ziegler Natta catalysts prepared industrially and in the laboratory?
- How are industrial polypropylene catalysts tested and the resultant polymer evaluated?
- What processes are used in the manufacture of polypropylene?
- What are biopolymer alternatives to polypropylene?

- What companies are the major industrial manufacturers of polypropylene?
- What is the environmental fate of polypropylene?

Terminology used in industrial polypropylene technology can
be baffling to the neophyte. This text will educate readers in the
jargon of the industry and demystify the chemistry of catalysts and
cocatalysts employed in the manufacture of polypropylene. Unlike
ethylene, propylene must be polymerized in a regioregular and
stereoregular manner to obtain the crystalline version of polypropylene ("isotactic") that has so many uses in the home, workplace
and in transportation.

Several techniques have been used to make the text "user
friendly." A thorough glossary is included as Appendix A. The glossary not only provides definitions of acronyms and abbreviations,
but also concisely defines terms commonly encountered in the
context of production and properties of polypropylene. An extensive index with liberal cross-referencing enables the reader to find
a subject quickly. Also, questions and exercises at the end of each
chapter allow the reader to assess whether he or she has mastered
the content. Answers are provided in Appendix B.

The following is an overview of the content and purpose of each
chapter:

- *Chapter 1 is used to recount the history of crystalline polypropylene and to describe basic properties and nomenclature
  for this versatile polymer. In addition, the most important
  industrial catalysts used for the manufacture of polypropylene are introduced. Also covered in Chapter 1 is an overview
  of stereochemistry, a crucial aspect that underpins properties and ultimately how polypropylene may be used in fabricated goods.*
- *Key polymer characterization methods in support of research
  and commercial production of polypropylene are discussed in
  Chapter 2.*
- *Features of catalysts and cocatalysts crucial to the manufacture
  of polypropylene are covered in Chapters 3–5. Chapter 3 dis-
  closes the origins of crystalline polypropylene, introduces key
  characteristics of Ziegler-Natta catalysts and includes an overview of mechanistic features of ZN polymerization. Chapter 4
  describes the various generations of industrial polypropylene*

catalysts. *Chapter 4 also reviews intermediate catalyst developments and the evolution of modern, supported high-activity catalysts that control both stereoregularity and particle morphology. Chapter 5 includes a description of aluminum alkyls, the organometallics that are absolutely essential to the functioning of industrial Ziegler-Natta catalysts. Chapter 5 also cites precautions that should be taken for the safe handling of hazardous aluminum alkyls, a topic of critical importance to all manufacturers of polypropylene.*

- *Chapter 6 is a discussion of single site catalysts, the most important of which are metallocene catalysts. Chapter 6 also covers the most common cocatalysts historically used with SSC and touches on a recent innovation wherein supported activators that require no cocatalysts are used.*
- *Chapter 7 describes the large-scale manufacture of polypropylene catalysts, including a description of the array of equipment required and the importance of recycle streams.*
- *Chapter 8 provides an introduction to the wide range of process technologies historically used to manufacture polypropylene and reviews industry trends, including the transition to "hybrid" processes.*
- *Chapter 9 illustrates synthetic procedures and safety precautions for the laboratory synthesis of 2nd and 4th generation polypropylene catalysts.*
- *Chapter 10 is concerned with laboratory polymerization testing protocols and procedures used in support of commercial catalysts.*
- *Chapter 11 surveys downstream aspects of polypropylene (additives, fabrication methods, environmental issues).*
- *Chapter 12 is a review of the global market, including a listing of the major producers of polypropylene that illustrates how the locus of manufacturing has shifted in recent years.*
- *Finally, Chapter 13 provides an assessment of the future of polypropylene, including a survey of key growth markets and an evaluation of emerging threats to the polypropylene supply chain. Chapter 13 also touches on new technologies developed for extraction of natural gas from shale, a positive development that could have huge long-range implications for the future of industrial polypropylene in the US. This is true because there are enormous shale deposits in the US and natural gas is a potential source of propylene monomer.*

This book should be considered complementary to *Introduction to Industrial Polyethylene* published jointly in 2010 by Scrivener Publishing, LLC and John Wiley & Sons, Inc. Because there are many commonalities between polyethylene and polypropylene, portions of text (especially in Chapters 3–5) have been reproduced with permission here. For example, aluminum alkyls function in generally the same way in Ziegler-Natta catalysts for polyethylene and polypropylene. Also, though there are subtleties having to do with stereochemistry and regiochemistry in propylene polymerization, essential features of the mechanism of Ziegler-Natta polymerization are very similar and textual similarities are inevitable.

The authors wish to thank former colleagues, friends and associates for their assistance in the production of this book. Thanks go to several former Akzo Nobel colleagues:

- William Summers and Jim Hatzfeld for their reviews of the information on the manufacturing and testing of industrial PP catalysts.
- Dr Biing Ming Su and Dr William Joyce for their comments on several sections, especially their help on analytical methods.

The authors would like to express gratitude to Drs Balaji Singh, Clifford Lee and JN Swamy of Chemical Marketing Resources, Inc. in Webster, TX for help on various aspects of the global polypropylene market. Thanks also to Dr Bill Beaulieu of Chevron Phillips Chemical, who provided a schematic depicting the Chevron Phillips "loop slurry" process for polypropylene.

Finally, the authors also greatly appreciate cogent input from the following:

- Dr Johst Burk, retired from AkzoNobel
- Dr Max McDaniel of Chevron Phillips Chemical
- Dr Roswell (Rick) King of BASF (formerly Ciba)
- Dr Rajen Patel of Dow Chemical

These men are accomplished professionals and have a combined total of more than 100 years' industrial experience. Each has expertise in one or more aspects of the polypropylene industry and each reviewed much or all of the manuscript and provided comments and suggestions for improvement. We are sincerely grateful for their help. However, we assume responsibility for any errors that may have survived the winnowing process.

With this book, we have endeavored to present essentials of the polypropylene industry in a way that would be readily grasped by chemists and engineers just beginning their journey in the fascinating universe of polypropylene. In the 1968 movie *The Graduate*, a man advised recent college graduate Benjamin to remember one word: "plastics." That one word could just as appropriately have been "polypropylene." Though few moviegoers would have known anything about polypropylene and script writers usually try to avoid polysyllabic words, few career choices at that time could have been as challenging, rewarding and professionally satisfying. We hope that the text will be a valuable reference for young graduates entering the polypropylene industry for years to come.

Dennis B. Malpass
Magnolia, Texas
March 21, 2012

Elliot I. Band
Pleasantville, New York
April 19, 2012

1

Introduction to Polymers of Propylene

1.1   Origins of Crystalline Polypropylene

Crystalline polypropylene (PP*) was unknown to the world before the 1950s. Though oligomeric and polymeric forms of propylene had been made before that time, they were typically amorphous, low molecular weight oils [1]. These liquid/oily polymers were produced by polymerization or oligomerization of propylene using free radical initiators or acidic/cationic catalysts at high temperature and pressure and were of marginal commercial value.

Giulio Natta is widely regarded as the discoverer of crystalline polypropylene, resulting from an experiment Natta performed in his lab in Milan, Italy on March 11, 1954 [2]. Without question, Natta contributed mightily to the fundamental understanding of crystalline polypropylene and its stereochemistry and richly deserved the admiration accorded him. However, with respect to US patent rights, who first prepared crystalline polypropylene remained a litigious issue for many years. Ownership of the US patent rights for crystalline

---

* Please see glossary for definition of abbreviations, acronyms and terms.

polypropylene was resolved only after nearly three decades of legal wrangling (interferences, appeals and much rancorous debate). Testimony in the case resulted in about 14,000 pages of text and 4,600 exhibits [3]. Nuances of polypropylene discovery and patent rights are beyond the purview of this text, but the interested reader can find more information in references [1–4]. In the final analysis, the courts awarded priority to Phillips Petroleum Company, tacitly acknowledging that Phillips chemists J. Paul Hogan and Robert L. Banks had first prepared crystalline polypropylene in an experiment conducted on June 5, 1951. On March 15, 1983, some 30 years after the original application, a definitive patent was finally issued to Hogan and Banks and rights assigned to Phillips covering solid polypropylene "having a substantial crystalline polypropylene content" [5].

In their discovery experiment, Hogan and Banks used a chromia-NiO catalyst supported on silica-alumina. Later, they demonstrated that it was the chromium portion of the catalyst that was responsible for polymerizing propylene to the crystalline polymer. Hogan and Banks were belatedly honored for their contributions to polyolefin technology with several awards, including the 1987 Perkin Medal by the Society of Chemical Industry. Hogan and Banks were also inducted into the National Inventors Hall of Fame in 2001 and the building in Bartlesville, Oklahoma in which they made their seminal discoveries on polyolefins was designated an "historic landmark" by the American Chemical Society [6]. Ironically, while it is true that the Phillips catalyst is today enormously important in manufacture of polyethylene (accounting for approximately a quarter of all polyethylene produced globally), it is unsatisfactory for commercial production of polypropylene [7].

In March of 1954, nearly 3 years after the Hogan-Banks discovery experiment, Natta synthesized crystalline polypropylene using a transition metal catalyst (and a metal alkyl cocatalyst) of the type that emerged from the remarkably fruitful work of Karl Ziegler in polyethylene. In recognition of the pioneering work of Karl Ziegler and Giulio Natta, polyolefin catalysts involving combinations of transition metal compounds with metal alkyl cocatalysts have become known as Ziegler-Natta catalysts [8]. Today, the vast majority (>97%) of industrial polypropylene as well as huge quantities of polyethylene are produced with modern versions of Ziegler-Natta catalysts. Ziegler and Natta shared the Nobel Prize in chemistry in 1963. A timeline of 20th century milestones in polypropylene is provided in Figure 1.1. More information on the origins and evolution of stereoregular polypropylene and Ziegler-Natta catalysts will be provided in Chapters 3 and 4.

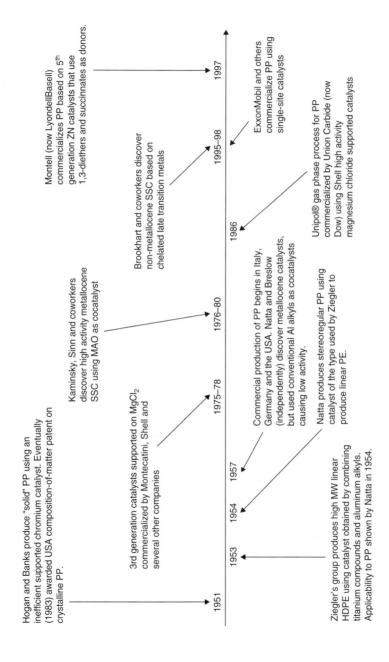

Figure 1.1 20th century milestones in polypropylene.

1.2   Basic Description of Polypropylene

Propylene (aka propene) has molecular formula C_3H_6. Other than ethylene, it is the simplest alkene, but in the parlance of the polypropylene industry is more commonly called an olefin. Propylene may be polymerized through the action of catalysts (eq 1.1). To obtain satisfactory quantities of stereoregular polypropylene, polymerization must be conducted under proper conditions using a transition metal catalyst and a metal alkyl cocatalyst. Other catalysts (free radical, cationic, *etc.*) typically produce low molecular weight amorphous polypropylene that is of limited commercial value.

$$n\ CH_2 = CHCH_3 \xrightarrow{\text{catalyst}} \sim (CH_2\overset{\overset{\displaystyle CH_3}{|}}{CH})_n \sim \qquad (1.1)$$

Note that the basic repeating unit of polypropylene contains a primary (1°), a secondary (2°) and a tertiary (3°) carbon atom. Note further that the tertiary carbon atom is chiral. Consequences of these characteristics will be developed in subsequent discussions. Ziegler-Natta catalysts are the most important transition metal catalysts for industrial polypropylene and titanium is, by far, the most widely used transition metal. We will introduce key aspects of Ziegler-Natta catalysts in section 1.5 below, and examine Ziegler-Natta catalysts in greater detail in Chapters 3 and 4.

Like Ziegler-Natta catalysts, single site catalysts (SSC) employ transition metal compounds and can be used to polymerize propylene. In the mid-1970s, Kaminsky and Sinn [9] found that SCC of extraordinarily high activity could be obtained if methylaluminoxanes (rather than conventional aluminum alkyls) are used as cocatalysts. This discovery sparked a renaissance in polyolefin catalyst research and a surge in literature and patents touting the attributes of polymers made with SSC. Single site catalysts exhibit exceptional activity and permit unprecedented control over the molecular architecture of polymers (see Chapter 6). However, it is important to keep SSC in perspective relative to industrial polypropylene. As of this writing, the quantity of commercial polypropylene produced with single site catalysts is very small (<3%). As more cost-effective SSC are developed, the percentage of polypropylene made with SSC will undoubtedly grow in the coming years.

However, Ziegler-Natta catalysts will remain the dominant catalysts for industrial polypropylene well into the 21st century.

Conditions for polymerization vary widely and polypropylene compositions also differ substantially in structure and properties. In eq 1, subscript n is termed the degree of polymerization (DP) and is greater than 1000 for most of the commercially available grades of polypropylene. As removed from industrial-scale reactors under ambient conditions, stereoregular polypropylene is typically a white powdery or granular solid with a density of ~0.90 g/cc. This density is significantly lower than most forms of polyethylene, which means that less weight of polypropylene is required to make an article relative to polyethylene.

Unlike ethylene polymerization, regiochemistry and stereochemistry possibilities exist in propylene polymerizations. Usually, propylene adds in a "head-to-tail" manner with Ziegler-Natta catalysts, but the reverse mode of addition (a "regioerror") is also possible. As noted above, the methine carbon atom in the polymer structure (eq 1) is chiral which creates a variety of stereoisomeric possibilities. The configuration of the methyl group in the polymer chain is indicative of what is called the polymer's "tacticity." If the methyl groups are predominantly oriented in the same direction, the polymer is designated "isotactic," a nomenclature derived from the Greek word for "same" or "ordered" proposed by Natta (following a suggestion from his wife who was a linguist) [10]. Isotactic polypropylene is by far the most common form of industrial polypropylene and contains substantial crystalline content. If the methyl group uniformly alternates from side to side along the polymer chain, the stereoisomeric form is termed "syndiotactic," and, like isotactic, also contains substantial crystalline content. If the methyl group is randomly oriented, the polymer is termed "atactic" and is a rubbery, amorphous, tacky material, generally considered to be undesirable. However, at least one company purposely manufactures atactic polymer, which has uses as an adhesive, among other applications. The three most common stereoisomeric forms are schematically illustrated in Figure 1.2. A variety of polymorphic forms and other tacticities are possible, but will not be discussed here. However, more information on stereochemistry and regiochemistry of polypropylene will be provided in Chapter 2.

Melting characteristics of various forms of polypropylene have been studied and results are not as straightforward as one might expect [11, 12]. Nevertheless, the melting point (T_m) of isotactic polypropylene is

- Isotactic PP (same orientation of methyl groups):

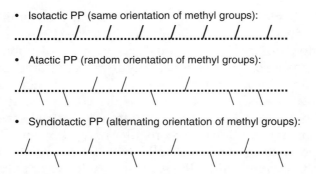

- Atactic PP (random orientation of methyl groups):

- Syndiotactic PP (alternating orientation of methyl groups):

Figure 1.2 Schematic representation of stereoisomers of polypropylene. where dashed lines represent the "backbone" of the polymer and solid lines indicate configuration of methyl groups along the chain.

typically reported to be in the range 160–170°C. T_m may be affected by a range of factors, including tacticity, molecular weight, and thermal history. Copolymers have lower T_m and lower crystallinity.

Though polyethylene is the least costly of the major synthetic polymers, polypropylene has higher T_m and is tougher (higher modulus and tensile strength) and can be used in applications where these attributes make polypropylene the material of choice. Like polyethylene, stereoregular polypropylene is a thermoplastic with excellent chemical resistance and toughness and can be processed in a variety of ways. Injection molding, fiber extrusion and film extrusion are fabrication methods that account for nearly 90% of all polypropylene applications. More information on fabrication methods will be provided in Chapter 11.

Because of the tertiary carbon atom, polypropylene is especially prone to attack by oxygen in ambient air, commonly termed "autoxidation" (see section 11.2 and Figure 11.1). To minimize oxidative degradation, the raw polymer is usually melted immediately after manufacture and an antioxidant is introduced. (Additives are essential to improve stability and enhance properties of polypropylene. See section 11.2) The molten product is shaped into translucent pellets and supplied in this form to processors (Figure 1.3). Pelletization increases resin bulk density resulting in more efficient packing and lower shipping costs. It also reduces the possibility of dust explosions while handling.

As noted above, polypropylene is a thermoplastic material. That is, it can be melted and shaped into a form which can then be subsequently remelted and shaped (recycled) into other forms.

Figure 1.3 Raw polypropylene resin is melted and shaped into pellets of the type shown above. Pelletization increases bulk density, improves handling characteristics and reduces shipping costs. Pellet size is typically ~3 mm (or ~0.1 in). (Reproduced from *Introduction to Industrial Polyethylene*, with permission of Scrivener Publishing LLC).

Propylene may be copolymerized with a range of other olefins, such as ethylene and α-olefins (1-butene, 1-hexene, etc.). The other olefins are termed comonomers and are incorporated into the growing polymer chain. Some types of vinylic comonomers cannot be used. For example, Ziegler-Natta catalysts are usually poisoned by oxygen-containing polar comonomers such as vinyl acetate. Consequently, copolymers of propylene and VA are not available from Ziegler-Natta catalysts. However, certain single site catalysts are more tolerant of oxygen-containing compounds. Hence, copolymers of propylene with polar comonomers may be available from single site catalysts (see Chapter 6) in the not-too-distant future. Such copolymers may have unique properties and applications.

Ethylene is the most commonly used comonomer. Some products also employ higher alpha olefins (1-butene, 1-hexene, etc.). Often, the pattern of incorporation of comonomer along the polymer chain is not statistical because of differences in reactivity (see discussion of kinetics and reactivity ratios in Chapter 8). This results in nonuniformities in content and distribution of comonomers along the polymer chain. This is called the composition distribution (CD) and may be determined by ^{13}C NMR analysis in combination with chromatographic methods discussed in Chapter 2.

Of course, regioerrors are not possible when ethylene is incorporated as comonomer, since carbons are equivalent. However, the polymer chain may contain "blocks" of ethylene units resulting from multiple insertions. This heterogeneity is termed "blockiness" and is represented schematically below.

$$CH_2 = CHCH_3 + CH_2 = CH_2 \xrightarrow{\text{catalyst}} \sim P\text{-}E\text{-}P\text{-}P\text{-}P\text{-}E\text{-}E\text{-}E\text{-}E\text{-}P\text{-}E\text{-}P\text{-}E\text{-}E\text{-}E\text{-}E\text{-}P\sim$$
$$\text{``P''} \qquad \text{``E''} \qquad\qquad\qquad \text{``Blocky copolymer''}$$

In addition to blockiness, alpha-olefin comonomers (1-butene, 1-hexene, etc.) may be incorporated in ways that result in regioerrors.

When propylene is copolymerized with large amounts (>25%) of ethylene, an elastomeric copolymer is produced, commonly known as ethylene-propylene rubber (EPR) or ethylene-propylene monomer (EPM) rubber. When a diene, such as dicyclopentadiene, is also included, a terpolymer known as ethylene-propylene-diene monomer (EPDM) rubber is obtained. EPR and EPDM may be produced with single site and Ziegler-Natta catalysts and are important in the automotive and construction industries. However, EPR and EPDM are produced in much smaller quantities (~1 million mt/y) relative to polypropylene. Moreover, EPR and EPDM are amorphous, while the most important type of industrial polypropylene ("isotactic") has substantial crystalline content. Accordingly, these elastomers are considered outside the scope of this text and will not be discussed further. However, random copolymers and impact copolymers are important products involving propylene and ethylene (at levels up to about 25%) and will be addressed in this text.

End groups of propylene polymers are most often saturated (simple alkyl groups). This is largely a consequence of using hydrogen as chain transfer agent. However, there are low levels of unsaturated sites owing to termination reactions by chain transfer *via* β-hydride elimination and hydride transfer to monomer. Termination reactions are addressed in the context of the mechanism of polymerization in Chapter 3.

Of course, there is abundant short chain branching (SCB) in polypropylene owing to the methyl group of propylene and other alpha olefins that are occasionally used. By convention, SCB implies 6 or fewer carbons. Long chain branching (LCB) in polypropylene is very low and is difficult to detect *via* the usual analytical methods [13].

Several grades of polypropylene are used in food packaging, *e.g.*, blown film for candy and snack foods. In Europe, Canada, Japan, the USA and other developed countries, the resin must satisfy governmental regulations for food contact. Catalyst residues are quite low in modern polypropylene and are considered to be part of the basic resin. In the USA, the resin (including additives; see Chapter 11) must be compliant with FDA requirements for food contact, such as hexane extractables. The procedure for registration of polymers with governmental agencies can be complicated and protracted. Normally, the registration specifies the permitted uses.

Polypropylene is available in a dizzying array of compositions, with different microstructures, various comonomers, a range of molecular weights, *etc.*, predicated by selection of catalyst, polymerization conditions and other process options. Since 1951, when a small quantity of a crystalline polymer was obtained unexpectedly from a laboratory experiment in Bartlesville Oklahoma, stereoregular polypropylene has grown enormously and is used today in megaton quantities in innumerable consumer applications. Though all forms of polyethylene (HDPE, LDPE, EVA, LLDPE, etc.) if taken together, remain the largest volume plastic, a recent analysis suggests that polypropylene is the *single* largest volume plastic produced globally, exceeding even that of HDPE which is the largest type of polyethylene manufactured. Global polypropylene production in 2010 was estimated to be about 48 million metric tons (~106 billion pounds) [14].

1.3   Types and Nomenclature of Polypropylene

The most important types of commercially available polypropylene are:

- homopolymer (HP)
- random copolymer (RACO, aka RCP)
- impact copolymer (also called heterophasic copolymer or HECO, aka ICP)

Key characteristics of each major type of polypropylene are summarized in Table 1.1. Figure 1.4 shows the approximate percentages of each type sold into the merchant market in 2008. The polymer produced in eq 1.1 is known as polypropylene or poly (propene).

Table 1.1 Characteristics of types of polypropylene.

Type of PP*	Abbreviation	Typical Olefin Comonomer	Range of Comonomer Content (wt %)	Impact Resistance	Film Clarity	Tensile Strength
Homopolymer	HP	none	0	poor	poor	good
Random Copolymer	RACO	ethylene	1–7	medium	good	medium
Heterophasic Copolymer (aka impact copolymer)	HECO (ICP)	ethylene	5–25	good	na**	poor

* Densities of HP, RACO and HECO are all in the range 0.89–0.91 g/mL

** Not applicable

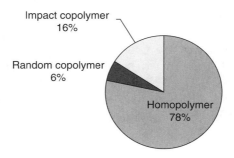

Figure 1.4 Types of industrial polypropylene in market in 2008. The total market in 2008 was estimated to be about 46 million metric tons (~101 billion pounds). Source: C. Lee and B. Singh, Chemical Marketing Research, Webster, TX, June, 2009.

Either is an acceptable name for homopolymer, according to IUPAC. However, polypropylene is by far the more commonly used name in industry and will be used exclusively in this text.

IUPAC nomenclature for copolymers is also not commonly used in industry. Since ethylene is the comonomer most often employed for RACO and HECO, the IUPAC name for such copolymers would be poly (propylene-*co*-ethylene).

Abbreviations for polypropylene are sometimes used to indicate the catalyst employed in its production or its stereochemistry. For example, polypropylene produced with metallocene catalysts (see Chapter 6) are often designated mPP. Isotactic and syndiotactic polypropylene are sometimes abbreviated iPP and sPP, respectively.

In addition to the nomenclature discussed above, manufacturers use their own registered trademarks. There are far too many trade names to include a comprehensive listing here, but examples are given below (along with the company that owns the trademark).

- Pro-fax (LyondellBasell)
- Vistamaxx (ExxonMobil)
- Achieve (ExxonMobil)
- VERSIFY (Dow)
- Fortilene (Solvay)
- Innovene (INEOS)
- INSPIRE (Braskem)
- Marlex (Chevron Phillips)
- Clyrell (LyondellBasell)
- Borstar (Borealis)
- TAFMER (Mitsui)

As is evident from Figure 1.4, more than three-quarters of the global industrial polypropylene market is homopolymer. Lesser amounts of copolymer are produced for specialized applications where specific attributes are desired, such as film clarity or superior impact resistance. While ethylene is most often used as comonomer, 1-butene and 1-hexene may also be used. Moreover, terpolymers are increasing in importance.

1.4   Molecular Weight of Polypropylene

Polypropylene manufacturers routinely supply data that correlate with molecular weight and molecular weight distribution. A measurement called the melt flow rate (MFR) is determined by the weight of polypropylene extruded over 10 minutes at 230°C through a standard die using a piston load of 2.16 kg. Reported in g/10 min or dg/min, melt flow rate is measured using an instrument called an *extrusion plastometer* (see Figure 1.5.) according to ASTM D 1238–04c Condition 230/2.16, where the latter numbers refer to the temperature and the load in kg on the piston of the plastometer, respectively. Melt flow rate is inversely proportional to molecular weight, *i.e.*, molecular weight decreases as MFR increases.

Figure 1.5 Melt flow rate of polypropylene may be measured on an instrument called an extrusion plastometer. (Photo courtesy of Zwick/Roell.)

The term melt index (MI) is sometimes applied (erroneously) to the low load MFR for polypropylene. ASTM suggests that MI (determined at 190°C) be reserved for polyethylene and MFR be used for all other plastics, regardless of conditions used (see note 27 on p 10 of ASTM D 1238–04c).

Another melt flow rate value for polypropylene may also be measured on the plastometer at 230°C, but under a load of 21.6 kg (ASTM D 1238–04c Condition 230/21.6). This MFR is also reported in g/10 min or dg/min. Dividing the MFR measured at high load by the MFR at low load affords what is called the flow rate ratio (FRR) as in eq 1.2. FRR is a dimensionless number which gives an

$$FRR = (MFR @ 21.6 \text{ kg}) \div (MFR @ 2.16 \text{ kg}) \qquad (1.2)$$

indication of breadth of the molecular weight distribution. As FRR increases, MWD broadens.

MFR and FRR measurements are inexpensive, relatively easy to conduct and are indicative of molecular weight and molecular weight distribution. Actual molecular weights may be determined using a variety of analytical methods, including gel permeation chromatography (GPC), viscometry, light scattering and colligative property measurements. (GPC is also called size exclusion chromatography or SEC.) However, these methods require more sophisticated instruments that are usually operated by highly trained technical personnel. Procedures are more costly and difficult to perform and do not lend themselves to routine quality control. Nevertheless, expressions having to do with molecular weight and molecular weight distribution of polypropylene obtained using these instrumental methods often appear in discussions of PP properties in patent and journal literature. The most important values are called the number average molecular weight $(\overline{M_n})$ and the weight average molecular weight $(\overline{M_w})$. The ratio $\overline{M_w}/\overline{M_n}$ is called the polydispersity index (PDI, also known as heterogeneity index and dispersity index) and is an indication of the broadness of molecular weight distribution. As polydispersity index increases, MWD broadens. Polydispersities typically range from 4–8 for polymer produced with Ziegler-Natta catalysts. However, polypropylene made with single site catalysts show polydispersities of 2–3 indicating a much narrower MWD.

The number average molecular weight $(\overline{M}_n)$ is calculated from the expression:

$$\overline{M}_n = \Sigma M_x N_x / \Sigma N_x \tag{1.3}$$

where M_x is the molecular weight of the xth component and N_x is the number of moles of the xth component. Weight average molecular weight $(\overline{M}_w)$ is calculated using the second order equation:

$$\overline{M}_w = \Sigma M_x^2 N_x / \Sigma M_x N_x \tag{1.4}$$

The third order equation provides the "z-average molecular weight" and is calculated from the expression:

$$\overline{M}_z = \Sigma M_x^3 N_x / \Sigma M_x^2 N_x \tag{1.5}$$

Higher order molecular weight averages may also be calculated, but are less important than $\overline{M}_w$, $\overline{M}_n$ and $\overline{M}_z$. For polydisperse polymers, such as polypropylene, the following relationship holds:

$$\overline{M}_z > \overline{M}_w > \overline{M}_n$$

Direct comparisons between melt flow rate and molecular weight of polypropylene should be made with caution. Such comparisons are appropriate only when the polymers have similar histories (made using the same catalyst, by the same process, etc.). An example of the relationship between melt flow rate and weight average molecular weight (determined from intrinsic viscosity) for polypropylene with similar histories is shown in Figure 1.6. However, polypropylene used in the latter study was produced using early generation ZN catalysts. An equation relating melt flow rate and molecular weight has also been published for "modern" polypropylene [15]:

$$\log MW = (-0.2773 \times \log MFR) + 5.7243$$

A graphical representation of the relationship between MFR and intrinsic viscosity for polypropylene prepared with modern ZN catalysts was shown by Del Luca, *et al.* [16]. Molecular weight measurement is discussed further in Chapter 2.

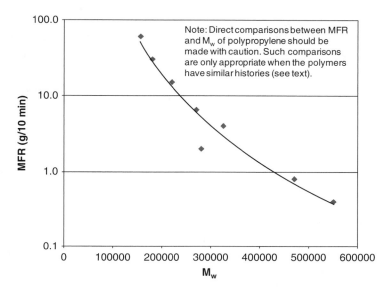

Figure 1.6 M_w vs. MFR for PP homopolymer. (from E. Vandenberg and B.C. Repka, *High Polymers*, John Wiley, 29, 365, 1977)

1.5   Transition Metal Catalysts for Propylene Polymerization

As previously mentioned, propylene may be polymerized by use of transition metal catalysts. To put these in perspective, one must recognize that transition metal catalysts have been used to produce virtually 100% of the cumulative trillion+ pounds of the global industrial output of polypropylene since commercial operations began in 1957. Transition metal catalysts are at the very heart of commercial processes used to manufacture polypropylene. Indeed, industrial polypropylene would be essentially unknown without transition metal catalysts.

In this section, we will introduce key characteristics of transition metal catalysts. From the standpoint of industrial polypropylene, Ziegler-Natta catalysts are by far the most important. Additional details on Ziegler-Natta catalysts will be provided in Chapters 3 and 4. As noted in section 1.2, the two most important catalysts are Ziegler-Natta and single site, both to be discussed in more detail in subsequent chapters. "Metallocene" catalysts (Chapter 6) may be considered to be a subset of what are more broadly designated as "single site catalysts." However, not all metallocenes are effective as catalysts for propylene polymerization and not all single site catalysts are metallocenes.

Polypropylene catalysts are most often produced using compounds from Groups 4–6 of the Periodic Table. Cocatalysts are required with all industrial versions of polypropylene catalysts. (Single site catalysts that do not require cocatalysts have been discovered, but are not yet used in industrial processes for polypropylene). In the vast majority of catalyst systems, alkylaluminum compounds (see Chapter 5) are used as cocatalysts. Modern Ziegler-Natta catalysts for polypropylene, are usually derived from inorganic titanium compounds and are most often supported on magnesium chloride. Industrial single-site catalysts (Chapter 6) commonly involve metallocene compounds of Zr, Hf or Ti. Non-metallocene single site catalysts based on late transition metals, especially Pd, Fe and Ni, began to emerge in the mid-1990s, though these are not industrially significant at this writing.

For industrial viability, transition metal catalysts for polypropylene must fulfill several key criteria:

- Activity must be high enough such that catalyst residues are sufficiently low in the final polymer to obviate post-reactor treatment to remove catalyst residues. As a rule of thumb, this requires that catalyst activity exceed about 150,000 lb of polypropylene per lb of transition metal.
- The catalyst must polymerize propylene with proper stereochemical control. In most industrial polypropylene processes, the desired stereoisomer is the isotactic version. Obtaining product with high isotactic content (usually >94%) makes it possible to forgo costly steps to remove atactic content.
- The catalyst must polymerize propylene with minimal "regioerrors." That is, polymerization must be selectively "head-to-tail" (see discussion in section 2.2.1).
- The catalyst must be capable of providing a range of polymer molecular weights. For Ziegler-Natta and single site catalysts, molecular weight is controlled primarily by use of hydrogen as chain transfer agent. Catalyst reactivity with hydrogen to control polymer molecular weight is called its "hydrogen response."

- Control of polydispersity is achieved primarily with the catalyst. Though each type of catalyst provides polypropylene with a characteristic range of molecular weight distributions, measures can be taken to expand marginally the range of achievable polydispersities. PDI are typically 4–8 for Ziegler-Natta catalysts and 2–3 for single site catalysts.
- If a RACO or HECO is the target resin, amounts and regiochemistry of comonomer incorporation must be acceptable. Quantities of comonomer, as well as the uniformity of incorporation (or lack thereof), are measures of whether satisfactory copolymerization has occurred. For example, a "blocky" copolymer is sometimes desired. ("Blockiness" of a copolymer can be determined by analysis of its composition distribution.)

Additional details will be provided in subsequent chapters on the composition and functioning of transition metal catalysts.

As noted above, transition metal catalysts are essential for the manufacture of polypropylene. It is not hyperbole to state that production of stereoregular polypropylene would not be possible without these catalysts. It is difficult to imagine a world without the huge number of products made from these versatile polymers in our homes, vehicles and workplaces. Ziegler-Natta will continue to be the dominant catalysts for the various forms of polypropylene well into the 21st century. However, as single site catalyst technologies mature, they will increase in importance and complement Ziegler-Natta catalysts in manufacture of polypropylene.

1.6   Questions

1. Who was the first to prepare crystalline polypropylene? When?
2. What percentage of industrial polypropylene is manufactured with Ziegler-Natta catalysts?
3. What are the three principal types of industrial polypropylene and what is the approximate percentage of each?
4. Why are copolymers of propylene with vinyl acetate not commercially available?

5. What is melt flow rate? How is it measured? What is it significance?
6. What is FRR? What is its significance?
7. What is the polydispersity index? What is its significance?
8. Provide key criteria for performance of industrial polypropylene catalysts.

References

1. RB Seymour, *Advances in Polyolefins*, Plenum Press, New York, 8, 1985.
2. N. Pasquini (editor), *Polypropylene Handbook*, Hanser (Munich), 9, 2005.
3. JP Hogan and RL Banks, *History of Polyolefins*, (R. Seymour and T. Cheng, editors), D. Reidel Publishing Co., Dordrecht, Holland, 105, 1985.
4. FM McMillan, *The Chain Straighteners*, MacMillan Publishing Company, London, 1979.
5. JP Hogan and RL Banks, US Patent 4,376,851, March 15, 1983.
6. MP McDaniel, *Handbook of Transition Metal Catalysts*, (R Hoff and R Mathers, editors), Wiley, 291, 2010.
7. *ibid*, 293.
8. J Boor, Jr., *Ziegler-Natta Catalysts and Polymerizations*, Academic Press, Inc., 34, 1979.
9. H. Sinn and W. Kaminsky, *Adv. Organometal. Chem.*, *18*, 99, 1980. See also W. Kaminsky, *History of Polyolefins*, (R. Seymour and T. Cheng, editors), D. Reidel Publishing Co., Dordrecht, Holland, 257, 1985.
10. FM McMillan, *The Chain Straighteners*, MacMillan Publishing Company, London, 127, 1979.
11. RA Phillips and MD Wolkowicz, *Polypropylene Handbook* (N. Pasquini, editor), Hanser (Munich), 160, 2005.
12. YV Kissin, *Alkene Polymerizations with Transition Metal Catalysts*, Elsevier, The Netherlands, 57, 2008.
13. T Jinghua, Y Wei, and Z Chixing, *Polymer*, 47, 7962, 2006.
14. H. Rappaport, *International Conference on Polyolefins*, Society of Plastics Engineers, Houston, TX, February, 2011.
15. M Dorini and G ten Berge, *Handbook of Petrochemicals Production Processes*, (RA Meyers, editor), McGraw-Hill (New York), 16.8, 2005; see also R Rinaldi and G ten Berge, *Handbook of Petrochemicals Production Processes*, (RA Meyers, editor), McGraw-Hill (New York), 16.26, 2005.
16. D Del Luca, D Malucelli, G Pellegatti and D Romanini, *Polypropylene Handbook* (N. Pasquini, editor), Hanser (Munich), 308, 2005.

2

Polymer Characterization

2.1 Introduction

There are many polypropylene characterization methods that are industrially important. These methods are used to characterize the whole range of homopolymers, random and block copolymers incorporating ethylene and other olefins, and terpolymers. Methods range from determination of basic physical properties, such as density, melting point and tensile strength, to measurement of applications-oriented properties, such as maximum continuous use strength and optical transparency. A summary of some of the important characterizations are shown in Table 2.1. Detailed descriptions of these and other polymer properties are available elsewhere [1].

This chapter describes characterization of several basic homopolymer properties that play a significant role in how polypropylene resin is processed and polymer properties that are achievable [2]. This includes two fundamental molecular properties and two physical properties of the raw resin. The former are polymer tacticity and polymer molecular weight, and the latter are polymer particle size distribution and bulk density. Measurement of these properties is discussed in subsequent sections.

Table 2.1 Selected characterization methods for polypropylene.

Polypropylene Property	Nature of Measurement	Why Important	ASTM Method(s)*
Brittle Temperature, °C	50% failure rate of a test specimen under specific impact	Limits of cold temperature service	D746
Coefficient of Thermal Expansion	Fractional change in length or volume per unit delta T by dilatometry	Dimensional stability, mold shrinkage	D696
Creep	Deformation under a load vs. time	Slow deformation under continuous stress	D2990
Crystallinity, %	Differential Scanning Calorimetry	Stiffness	D3418
Density, g/mL	Displacement	Weight of a given form	D1505
Enthalpy of Fusion, kJ/mole	Differential Scanning Calorimetry	Measure of crystallinity	D3418
Flexural Modulus, (GPa)	Load at yield for a molded piece held across two supports and subjected to a vertical load	Stiffness, resistance to bending or deformation under stress	D790
Glass Transition Temp, °C	Differential scanning Calorimetry	Embrittlement temperature	D3418

(Continued)

Table 2.1 (cont.) Selected characterization methods for polypropylene.

Polypropylene Property	Nature of Measurement	Why Important	ASTM Method(s)*
Hardness, Rockwell Hardness Number	Depth of impression from an indenter vs. load	Resistance to scratches	D785, D2583
Haze, %	Light beam transmittance and scattering	Visual clarity	D1003
Heat Deflection Temp., °C	Test bar deflection at temperature under a given load	Limit of hot temperature service	D248
Intrinsic Viscosity, dl/g	Viscosity per unit concentration extrapolated to infinite dilution	Related to MW; may be indicative of processability	D2857, D446
Isotactic Index, wt %	Physical extraction methods or NMR	Attainable crystallinity, Extractables	D5227 (hexane)
Max. Continuous Use Temp., °C	50% loss in tensile strength at 100,000 hrs	Service life	Underwriter Laboratories
Melt Flow Rate, dg/min	Flow through a capillary rheometer under a given load	Related (inversely proportional) to MW. Indicative of Processability	D1238
Melting Point, °C	Differential scanning Calorimetry	Service temperature limit	D3418

(Continued)

Table 2.1 (cont.) Selected characterization methods for polypropylene.

Polypropylene Property	Nature of Measurement	Why Important	ASTM Method(s)*
Mold Shrinkage, %	Dimension change of a molded piece vs. mold dimensions	Mold design	D955
Molecular Weight	Gel permeation chromatography	Melt strength	D6474
Molecular Weight Distribution	Gel permeation chromatography	Melt processing characteristics	D6474
Notched Izod Impact at 23°C, kJ/m	Kinetic energy to break a notched specimen with a swinging pendulum	Impact strength	D256
Oxygen Permeability, Barrers	Oxygen passing per unit area through a unit thickness per time for a given pressure, as measured by a sensor	Food preservation	F2622
Particle Size Distribution, wt %	Mechanical classification of virgin resin	Resin processing and transport	D2124
Resistivity, ohm-cm	Electrical potential applied to opposite faces of a test block	Usage in insulator applications	D2305, D257

(Continued)

Table 2.1 (cont.) Selected characterization methods for polypropylene.

Polypropylene Property	Nature of Measurement	Why Important	ASTM Method(s)*
Specular Gloss at 20°C, %	Reflectance at specified angles	Surface shine	D523, D2457
Tensile Strength, (Mpa)	Stretching a molded sample	Load limit	D638
Transparency	Light transmittance through a test sheet	Visual clarity	D1746
Vicat Softening Temperature, °C	Flat ended needle penetration to 1 mm depth	Heat softening properties	D1525
Water Absorption	Immersion	Effect on electrical insulation and dimensional integrity	D570
Xylene Insolubles, wt%	Precipitation	Attainable crystallinity, Extractables	D5492

*Other ASTM methods may also apply.

2.2   Polymer Tacticity

2.2.1   Introduction

Polypropylene's key properties as an inexpensive, lightweight engineering plastic are its tensile strength and stiffness. Crucial to this is its ability to crystallize. This confers strength, permitting the resin to be useful in molded forms. Crystallinity requires a regular structural repeat unit. In order to achieve this, there are two requirements. First, the propylene monomer units must be regiospecifically enchained in a regular head-to-tail manner (Figure 2.1).

With most polymerization catalyst systems this criterion is met; head-to-head and tail-to-tail enchainment is almost entirely avoided [3, 4].

Secondly, because the carbons are tetrahedrally coordinated and form a three dimensional structure, the methyl group on each monomer must be oriented the same way with respect to the polymeric chain backbone. This is illustrated in Figure 2.2 which shows four structural categories which are of industrial significance. The stereochemistry is described in terms of the spatial relationship of adjacent methyl groups. They may have either the same stereochemical handedness, which is termed "m" (from meso), or opposite placement, which is termed "r" (from racemic). Figure 2.2 demonstrates this.

The first, and most important, category (Figure 2.2(a)) is isotactic polymer, where the methyl groups are all oriented in the same manner. This comprises most of the world's industrial production. The second type is syndiotactic (Figure 2.2(b)), where the methyl groups alternate in their orientation with respect to the backbone. Syndiotactic polymer has some industrial importance. It is produced by certain metallocene catalysts and is discussed in Chapter 6.

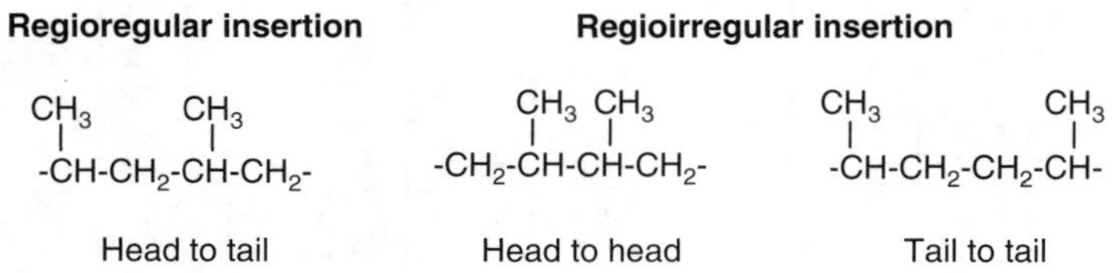

Regioregular insertion   **Regioirregular insertion**

CH_3 CH_3 CH_3 CH_3 CH_3 CH_3
| | | | | |
$-CH-CH_2-CH-CH_2-$ $-CH_2-CH-CH-CH_2-$ $-CH-CH_2-CH_2-CH-$

Head to tail   Head to head   Tail to tail

Figure 2.1 Enchainment of propylene monomers.

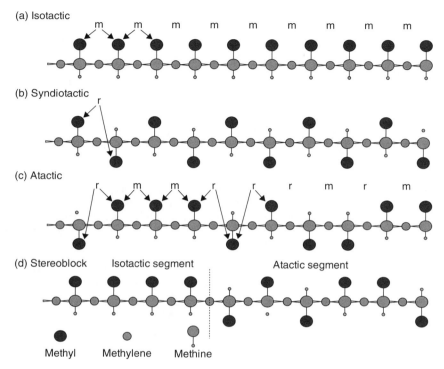

(a) Isotactic

(b) Syndiotactic

(c) Atactic

(d) Stereoblock   Isotactic segment   Atactic segment

Methyl   Methylene   Methine

Figure 2.2 Four important tacticities of regioregular polypropylene. The placement of the methyl groups (-CH$_3$) defines the polymer tacticity. Methylenes (-CH$_2$-) are shown smaller to emphasize their stereochemical placement below the plane of the paper. The methine carbons(CH) lie in the plane of the paper are shown with the associated proton to emphasize the tetrahedral bonding. Methyl groups sit above the plane of the paper.

The third type is an atactic structure (Figure 2.2(c)), in which the placement of methyl groups follows no repeat pattern. Atactic polymer is significant in the production of adhesives, and it is discussed in Chapter 4. The fourth type, (Figure 2.2(d)), is a stereoblock structure, where a section of the polymer has an isotactic placement of methyl groups, followed by a section or block with random or atactic methyl group placement. This is important for impact grades of polypropylene, which are produced by multi-reactor processes discussed in Chapter 8. Other structural arrangements can be imagined, at least from an academic perspective.

Isotactic stereochemistry is the centerpiece of industrial polypropylene. The methine carbon of each enchained monomer is chiral. That is, in principle, each of the four bonds is distinct. As a result, each

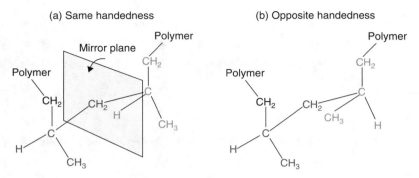

Figure 2.3 Chiral relationship of a polypropylene diad. Two propylene monomers (a "diad") are shown where (a) is an "m" diad, and (b) is an "r" diad. Rotation about the C-CH$_2$-C bonds will not convert (b) into (a). Each methine carbon is chiral.

methine has a handedness, or chirality. This is shown in Figure 2.3 for a pair of enchained propylene monomers in the polymer. In the same way that the right hand and left hand are mirror images of each other, so, too are the two possible stereoplacements of groups about the methine. Thus, isotactic polymer represents a repeated, consistent handedness at the methine carbon. This stereoregular repeat unit permits polypropylene to crystallize.

The propylene monomer itself is not chiral, but propylene is termed "prochiral". The methine carbon in propylene is in a 3-coordinate, planar environment (sp^2 hybridization), and it becomes tetrahedral (sp^3 hybridization) on enchainment. This transition develops opposing handedness in the methine carbon depending upon which of the two propylene faces adds to the growing chain. Hence, propylene is prochiral because each methine carbon will become chiral once inserted.

How does the polymerization process consistently control chirality as each monomer is inserted? Based on the experimental evidence, this control, *i.e.* a consistent stereochemical insertion pattern into the growing chain, arises directly from the stereochemistry of the catalyst site. Its chirality induces a specific handedness in the insertion step. This is described in section 3.7 [5]. It may seem counterintuitive, but a perfectly isotactic polymer chain is not a chiral macromolecule, as such, as shown by the mirror plane in Figure 2.3. That is, isotactic polypropylene, while demonstrating a consistent handedness or chirality at individual methine sites on the chain, is not chiral on a macroscale.

Stereoregular tacticity, first discussed in section 1.2, is not synonymous with crystallinity. Rather, the former is a prerequisite for the latter. Normally, crystallization takes the form of a helical chain structure which then folds itself into lamellae, about 50–200Å thick [6]. In practice, most polypropylene is semicrystalline due to occasional errors in tacticity and due to the difficulty of orienting long polymer chains into crystallites. As molecular weight increases or as the cooling rate of the polymer melt increases the crystallinity decreases. In commercial isotactic polypropylene up to about 70% crystallinity is typically achievable [7].

In commercial reality, the regio- and stereoregular enchainment of propylene monomers is not perfect, and so the evolution of Ziegler Natta propylene polymerization catalysts is, in large part, a history of improving the stereochemistry of the enchainment process. As the atactic content of the polymer increases, physical properties change. Stiffness, strength, and melting temperature decrease. Clarity, impact strength, and ductility increase. Solvent extractables also increases, which impacts allowable food contact uses. Thus, assessment of tacticity is critical to understanding polymer behavior.

There are two principal approaches to assessing polypropylene stereoregularity: microstructural measurements and physical measurements. The former is almost universally measured by high resolution ^{13}C nuclear magnetic resonance (NMR) spectroscopy and is discussed in the next section. The latter, which will be discussed later in this chapter, consists of two principal analyses: isotactic index and xylene insolubles.

2.2.2 Measurement of Polymer Microtacticity by ^{13}C NMR

^{13}C NMR spectroscopy is a powerful instrumental method that measures the short range structural environment ("microtacticity") for carbon atoms. It relies upon small changes in the resonance frequency in the nuclear magnetic spin of ^{13}C in a strong magnetic field, where the specific field resonance frequency is influenced by the local chemical environment about the carbon atom. The resonance frequency is called a chemical shift, when referenced to a standard. Sophisticated and expensive instrumentation is required. Experimental precision relative to a standard can be achieved to 4 significant figures.

Historically, ^{13}C NMR was developed in the same period as polypropylene technology, with the first ^{13}C NMR spectra recorded in 1957 [8]. It is now the undisputed primary standard for probing polypropylene microstructure. The ^{13}C NMR measurement is performed using a sample tube containing polypropylene typically dissolved in chlorinated and deuterated hydrocarbon solvent at 110–140°C. The chlorination helps dissolve the polymer, and the deuteration is used as a signal lock. Once inserted into the NMR probe in the magnetic field, the sample tube is repeatedly pulsed with a radio frequency signal, the free induction decay signal is acquired, and it is converted by a Fourier transform into the ^{13}C spectrum. Signal acquisition takes some hours. For an in-depth description of the NMR technique, the reader is referred to other references [9].

Using ^{13}C NMR, the microstructure of polypropylene is probed by the chemical environment of the pendant methyl groups. ^{13}C NMR is not typically used as a routine quality control method for PP because it requires sophisticated, expensive instrumentation and highly trained technical specialists. Recall from section 2.2.1 and Figure 2.2 that adjacent methyls may have either the same handedness, termed "m" or the opposite, termed "r". Extending this to the next pair of methyl groups the same relationships can be established. Under routine conditions a high resolution ^{13}C NMR instrument is useful for resolving the small shifts to the chemical environment of each methyl from the two adjacent methyl groups on each side [10]. That is, for each methyl group the particular arrangement of the two adjacent methyl groups on each side gives rise to a specific, resolvable chemical shift. This is called a pentad structure, for the five methyl groups involved, and it is denoted by a four-letter sequence. Each letter represents the stereochemical relationship of two adjacent methyls. Returning to the idealized structures in Figure 2.2, perfect isotactic polymer is characterized by a single NMR resonance corresponding to an mmmm pentad; each pair of methyl groups in the pentad has the same stereochemical configuration. An occasional error produces an r diad. For a catalyst that produces mainly isotactic polymer this error will be followed by correct stereochemical insertion of the next monomer, and an rr triad results, as shown in Figure 2.4.

Syndiotactic polypropylene is characterized by one NMR resonance corresponding to an rrrr pentad; each pair of methyl

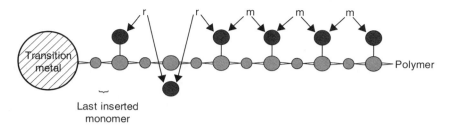

Figure 2.4 Single insertion error in isotactic chain propagation. A single insertion error in the isotactic chain results in an rr triad.

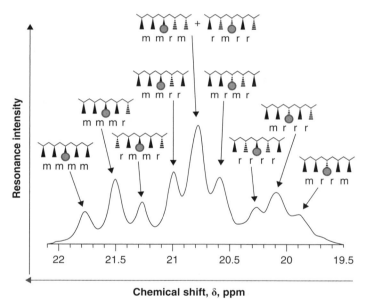

Figure 2.5 Pentad assignments for methyl groups in the ^{13}C NMR spectrum of atactic polypropylene. Adapted from http:// bama.ua.edu/~kshaughn/ch338/ handouts/poly-lecture.ppt Methyl group corresponding to each spectral line is shown by circle. The chemical shift of each ^{13}C methyl resonance is shown by the x-axis, relative to the methyl shift of tetramethylsilane internal standard.

groups in the pentad has opposite stereoconfigurations. By contrast, atactic polypropylene is characterized by nine NMR signals whose intensities correspond to a statistical distribution of the possible pentad structures. Figure 2.5 shows a representative ^{13}C spectrum of the polypropylene methyl region with assignments.

2.2.3   Total Isotactic Index

2.2.3.1   Introduction

From a physical standpoint, isotactic polymer is virtually insoluble in organic solvents compared to atactic polymer, a property recognized in the early work of Natta [11]. Indeed, the earliest industrial processes relied upon this behavior for process design. Thus, slurry processes (see Chapter 8) were commercialized which utilized an inert hydrocarbon polymerization medium. This dissolved much of the atactic inevitably formed from the early $TiCl_3$ catalysts. Such processes used hexane, heptanes, isooctane, and kerosene as diluents/solvents [12]. Accordingly, a laboratory physical test was developed along the same lines to measure the polymer which could be solubilized by the hydrocarbon solvent. The test described in this section uses heptane, which is generally accepted as the solvent of choice [13]. The polymer sample is extracted in a Soxhlet apparatus, as shown in Figure 2.6 and the solubilized polymer is quantified as a fraction of the total polymer.

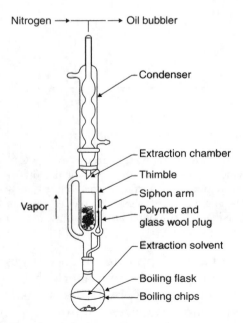

Figure 2.6 Soxhlet extraction apparatus [16]. Adapted from http://technologylodging.com/wp-content/uploads/2010/05/SoxhletExtractor.gif.

Table 2.2 Successive extractions of polypropylene by different solvents [14].

Sample Number (Source)	Soxhlet Extraction Method	Extracted (wt%)	Cumulative (wt%)	Nature of soluble polymer from pentad and M_w analyses
1 (Virgin Polymer)	–	0	–	mmmm = 92.5% M_w = 396 K
2 (1, Extracted)	Heptane	4.1	4.1	mmmm = 20.2% M_w = 25 K
3 (Insolubles from 2, Extracted)	Octane	3.9	8.0	mmmm = 78.7% M_w = 84 K
4 (Insolubles from 3, Extracted)	Xylene	2.9	10.9	mmmm = 90.6% M_w = 200 K
5 Remaining insoluble polymer	–	89.1	100	mmmm = 96.5% M_w = 481 K

Table 2.2 Data assembled from R. Paukkeri and A. Lehtinen, *Polymer*, vol 35, number 8 1673–1679, 1994.

Extraction with various solvents results in different solubilized fractions. Table 2.2 summarizes reported data for a polypropylene sample prepared with a supported catalyst in the gas phase [14].

The data demonstrate the empirical nature of the test method; extraction with boiling octane solubilizes more polymer than boiling heptanes, and xylene extraction even more. The heptanes extractables contain an appreciable fraction of mmmm material of low molecular weight, *i.e.* short chain isotactic polymer. Octane extractables have an even higher fraction of mmmm pentads, and xylene extractable material is highly isotactic. Therefore, Table 2.2

shows heptane is a more selective extractant than octane for atactic material. Table 2.2 also shows that xylene is not suitable as a Soxhlet extractant because it extracts highly isotactic polymer.

2.2.3.2   Measurement Method

Equipment Requirements

- Soxhlet extractor apparatus (Figure 2.6) with 500 mL round bottom flask and condenser
- Porous glass fiber thimble
- Heating mantle and temperature controller
- Cryomill for polymer samples containing large particles

Analytical Procedure

1. Tare a glass fiber thimble and a piece of glass wool to be used as a plug.
2. If the polymer originates from polymerization in a hydrocarbon slurry, a 10.0% aliquot of the diluent is taken from the filtration of the polymer. It is evaporated in a tared pan to give the soluble polymer (SP). Calculate the total soluble polymer (TSP):

$$TSP(g) = 10^* \, SP$$

   If the polymer originates from a liquid propylene polymerization this step is omitted.
3. Oven-dry the starting polymer sample to remove any solvents or moisture. Obtain the net weight of dry polymer (DP).
4. Examine the polymer; if it consists of a fine powder, it may be used as is. If it is large particles, containing particles of 2 mm diameter or above, then cryomill the polymer first with dry ice chips (Figure 2.7) [15]. and redry it.
5. Add 10.0 g of the polymer sample to the thimble, add the glass wool plug and reweigh.
6. Place the thimble into the Soxhlet extractor (Figure 2.5).
7. Add 300 mL heptane and some boiling chips to the 500-mL flask and assemble the apparatus.

Figure 2.7 Mill Suitable for reducing polymer particle size [17]. Source: Paul N. Gardner Co, Inc., http://www.gardco.com/pages/dispersion/ml/wileyminimill.cfm Polymer and dry ice are introduced into the hopper. Ground materials exit through a screen just above the receiving jar.

8. Purge the apparatus with nitrogen for a few minutes, and leave a slow nitrogen stream going to the bubbler which is attached to the apparatus by a T connection.
9. Apply heat to bring the heptanes solution to a reflux, turn on the condenser water and extract the polymer for three hours.
10. Cool and dry the thimble in a forced air oven overnight. Measure the net weight of polymer in the thimble. Calculate the total extractable polymer (TEP):

TEP (g) = (10.0 - final net weight)*(DP/10.0)

11. Weigh the thimble and calculate the total isotactic index (TII):

TII (wt%) =100*[1-(TSP +TEP)/TP]

Where TP (g) = TSP + DP

2.2.4   Total Xylene Insolubles

2.2.4.1   Introduction

Complementary to the total isotactic index, which is an extraction technique, is the test for xylene insolubles, which is a total dissolution and reprecipitation technique. Both isotactic and atactic polymer dissolve in hot xylene, and by cooling slowly the isotactic fraction can be cleanly recrystallized, leaving atactic polymer in solution. Therefore, total xylene solubles is a somewhat more accurate measure of atactic polymer than heptane extractables, because the latter also extracts some low MW isotactic polymer. Further, because this test dissolves the entire polymer sample, it does not require cryomilling the polymer, as is needed for the isotactic index measurement of large polymer particles. So, in most cases measurement of xylene solubles is preferred, and the results have been found to correlate with film and fiber application characteristics [18]. A more detailed procedure may be found in ASTM D5492, and automated instrumentation is also available [19].

The isotactic pentad fraction, mmmm, has been found to correlate with xylene insolubles. A representative correlation, drawn from two references is shown in Figure 2.8 [20, 21]. Sometimes lower correlations have been reported [22].

Alternatives have been developed to correlate total xylene solubles with other measurement methods, among these FTIR and low resolution broad band NMR. The latter is of particular interest because it requires modest equipment, minimal operator training and sample preparation, and provides a rapid response. Utilizing the polymer powder directly as produced, it is attractive for a

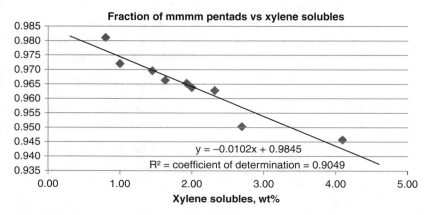

Figure 2.8 Correlation of pentad isotacticity and xylene solubles. Data assembled from 1) C. Meverden, A. Schnitgen, S. Kim, US patent 7465776, Bulk polymerization using supported catalyst, producing a high MFR polymer, and 2) V. Virkkunen, Polymerization of Propylene with Heterogeneous Catalyst, Active Sites and Corresponding Polypropylene Structures, Academic Dissertation, U. of Helsinki, 2005.

quality control environment [23]. This technique is not to be confused with high field ^{13}C NMR, although the same basic principles
are involved. Pulsed broad band NMR utilizes the relatively large
difference in the free induction decay (FID) times between the proton signals from crystalline, *i.e.* isotactic, and amorphous, *i.e.* atactic
polymer domains. For a given type of polypropylene the FID signal
is broken down by an algorithm into a sum of separate decay time
functions. These are quantified by calibration with the classical wet
method, described above, which serves as the primary standard.
The result is typically a good linear correlation curve which is used
to analyze subsequent related polymer samples. Commercial bench
top instruments are available [24].

2.2.4.2   Measurement Method

Equipment and Chemical Requirements

- Oil bath, immersion heater, thermocouple and stir bar
- Temperature controller
- Magnetic stirrer and stir bar
- Thermostated cooling bath
- 500-mL Erlenmeyer with condenser
- Funnel, filter flask and filter paper
- Steam plate
- Ortho-xylene, low residue

Analytical Procedure

1. If the polymer originates from polymerization in a
   hydrocarbon slurry, a 10.0% aliquot of the diluent is
   taken from the filtration of the polymer. It is evaporated in a tared pan to give the soluble polymer (SP).
   Calculate the total soluble polymer (TSP):

$$TSP(g) = 10^* SP$$

   If the polymer originates from a liquid propylene polymerization this step is omitted.
   Oven dry the starting polymer sample to remove any solvents
   or moisture. Obtain the net weight of dry polymer (TDP).
2. Accurately weigh 3.00 g polymer into a tared Erlenmeyer.

3. Add 250.0 g of xylene, and reweigh the flask.
4. Add the condenser, a magnetic stirring bar and attach to the nitrogen supply with a T connection to a bubbler.
5. Begin a nitrogen flow through the bubbler and heat the oil bath to 135°C with magnetic stirring.
6. Maintain at 135°C until the sample dissolves entirely to give a homogeneous solution.
7. Remove the Erlenmeyer from the oil bath and cool on the bench while maintaining a gentle nitrogen flow through the bubbler.
8. When the flask is warm to the touch place it in a thermostatted bath at 22°C and let the mixture cool completely. During this time isotactic polymer will precipitate. Keep flask at 22°C for two hours.
9. Filter the mixture by gravity through a filter paper, collecting about one half of the filtrate.
10. Weigh the filtrate into a tared disposable aluminum pan. Let FW= filtrate weight.
11. Evaporate the filtrate inside a hood on a steam plate. A gentle flow of nitrogen directed into the pan will assist the evaporation. Measure the net weight of xylene soluble polymer (XP).
12. Evaporate a blank of xylene = FW in similar way. Let the net residue =B.
13. Calculate the total xylene soluble polymer (TXP):

$$TXP = [(XP-B)* 250/FW] * TDP/3.00$$

14. Calculate the total xylene insolubles TXI).

$$TXI (wt\%) =100*[1-(TSP +TXP)/TP]$$

Where TP (g) = TSP + TDP

2.3   Molecular Weight and Molecular Weight Distribution

2.3.1   Introduction

The importance of polypropylene stereochemistry has been discussed above. The second major factor in determining polypropylene's utility is its degree of polymerization. This is measured as the molecular weight and the molecular weight distribution. This is important for both processing and final product performance. The molecular weight and the distribution are functions of both the catalyst and the polymerization conditions, especially the hydrogen concentration during polymerization. For processing, polypropylene must be melted in order to be converted into finished products. The processability characteristics are directly related to the melt viscosity, and that, in turn, is a function of the polymer chain molecular weight and the molecular weight distribution of the assembly of polymer chains. For example, lower molecular weight reduces melt viscosity, making it easier to inject or extrude molten polymer, but at the same time the melt strength of the polymer may decrease. Similarly, the performance characteristics of the final formed polymer product are affected by the molecular weight. For example, higher molecular weight increases the interlamellar entanglements, building impact strength. However, higher molecular weight can reduce crystallinity, which reduces hardness. Thus, molecular weight is the second principal factor in determining how a polymer will be processed and whether it will have the final required performance properties [25]. The following sections describe methods to assess the molecular weight and molecular weight distribution by chromatography, viscometry and capillary rheometry.

2.3.2 Gel Permeation Chromatography

2.3.2.1 Introduction

From an analytical perspective, the measurement of polymer molecular weight and molecular weight distribution is best done by high temperature gel permeation chromatography (HT-GPC) which also provides detailed information on the modality of the molecular weight distribution. GPC separates polymer macromolecules based on a size exclusion principle. Polymer macromolecules penetrate a macroporous crosslinked gel based on size. The largest molecules are excluded to the greatest extent, and thus elute first. The analysis is performed using a specialized high temperature high performance liquid chromatograph. See Figure 2.9.

2.3.2.2 Measurement Method

The chromatographic procedure will be described briefly. Polypropylene analysis requires a high temperature GPC instrument.

Figure 2.9 Example of an HT GPC instrument: Viscotek system. Reproduced by permission of Malvern Instruments, Inc. Source: http://www.malvern.com/LabEng/products/viscotek/viscotek_gpc_sec/viscotek_ht_gpc/ Viscotek_ht_gpc_polyolefin_analyzer.htm. The Viscotek® HT GPC system includes, from left to right, heated, automated sampler, high temperature module for the GPC columns and all detectors, solvent degasser, isocratic solvent pump, and computer/software.

Analysis starts with dissolution of the polymer in a hot solvent such as trichlorobenzene to form a dilute solution, 0.1–1.0 g/liter at 140°C. Then the solution is injected into the chromatograph and eluted with trichlorobenzene at 140°C through a chromatographic column containing a solid crosslinked macroporous gel, such as styrene-divinylbenzene. The gel separates polymer molecules by a size exclusion principle. For this reason the technique is also called size exclusion chromatography (SEC). The eluted polymer typically forms a continuous distribution curve that is detected by a refractive index detector and calibrated against known standards [26]. The output is in the form of a chromatogram, which plots signal intensity (*i.e.* polymer concentration) vs eluant time. From the shape of that chromatogram one can calculate different types of average molecular weights and the breadth or spread of the molecular weights, *i.e.*, the molecular weight distribution. Additionally equipped with certain detectors, such as low angle light scattering and viscometry detectors, HT-GPC is capable of providing a wealth of polymer information, including absolute molecular weight, molecular size, chain branching, and intrinsic viscosity analyses. The discussion that follows is limited to molecular weight information.

The most widely used are the number average molecular weight and the weight average molecular weight, represented as M_n and

M_w, respectively, introduced previously in section 1.4. They are defined as follows:

$$M_n = \frac{\Sigma_i^0 M_i N_i}{\Sigma_i^0 N_i}$$

$$M_w = \frac{\Sigma_i^0 M_i^2 N_i}{\Sigma_i^0 N_i}$$

where M_i is the molecular weight of the ith fraction, and N_i is the number of chains of molecular weight M_i.

Experimentally, M_n and M_w are determined as

$$M_n = \frac{\Sigma_i^0 A_i}{\Sigma_i^0 A_i/M_i}$$

$$M_w = \frac{\Sigma_i^0 A_i M_i}{\Sigma_i^0 A_i}$$

where A_i is the signal area of the ith fraction, and M_i is the molecular weight of the ith fraction.

M_n is understood intuitively as the total weight of polymer divided by the number of polymer chains. M_w is always larger than M_n; it gives more weight to the longer polymer chains. It can be understood in terms of the probability of finding a monomer unit. For a given set of polymer chains, it is more probable to find a given monomer unit in a long polymer chain than in a short polymer chain, simply because there are more monomers in a long chain. Thus, M_w squares the weight of each polymer chain to account for this.

Finally, the polydispersity is defined as PDI = M_w/M_n. This provides a measure of the breadth of the polymer distribution. A monodisperse polymer, *i.e.* all polymers of the same length, has PDI = 1.0. Typical polypropylene PDIs = 4–8 for Ziegler Natta catalysts. Metallocene catalysts are capable of producing resins with PDI = 2–3, as are controlled rheology grades (see section 11.3) produced by post-reactor treatments. An example of a GPC chromatogram for a multimodal polypropylene is shown in Figure 2.10.

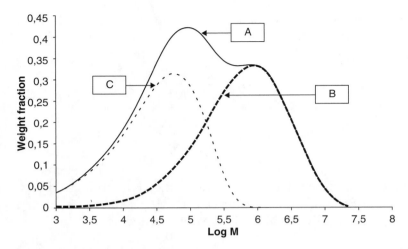

Figure 2.10 HT GPC chromatogram of a bimodal isotactic polypropylene [27]. The weight fraction on the Y axis is derived from the signal intensity. The logarithm of the molecular weight on the X axis is derived from the eluant retention time. The retention time for lower molecular weight polymer is longer than the retention time for high molecular weight polymer. The sample, curve A, is deconvoluted to two components B and C that were physically mixed to produce sample A. M_w = 671,000, M_n = 27,000, PDI = 24.9. Source: M. Dupire, J. Michel, US patent 6723795.

2.3.2.3   Temperature Rising Elution Fractionation

For the isolation of discrete molecular weight fractions, a related chromatographic technique is temperature rising elution fractionation (TREF) [28]. It is particularly useful for investigating the complex composition distribution of propylene copolymers. TREF is a two step process. In the first step polymer dissolved in a hot *o*-xylene solution is slowly precipitated by controlled cooling over many hours onto an inert support such as glass beads so that the higher isotactic fractions, *i.e.* more readily crystallizable chains, precipitate first, and then lower tacticity fractions. A typical precipitation might involve cooling the solution from 130 C to 25 C at <0.1 C/min. During the cooling crystal segregation takes place. In the second step, the coated support is loaded into a thermostatted column. The column is then eluted with a solvent, such as xylene, under a stepwise series of increasing temperature conditions. During this phase, polymer fractions sequentially dissolve, with the least crystalline dissolving first. The polymer in each eluant fraction is isolated by reprecipitation with acetone and then each fraction is analyzed by HT-GPC, DSC, NMR or other techniques. TREF requires careful attention to both the experimental apparatus construction and experimental technique, and it is time-consuming. Various references describe HT-GPC and TREF in detail [29].

2.3.3   Intrinsic Viscosity

2.3.3.1   Introduction

As the reader will surmise, HT-GPC and TREF are excellent analytical techniques, but require expensive instrumentation and an analytical specialist. Alternative analytical molecular weight methods are available that require less sophisticated instrumentation and that, to a certain extent, can be correlated with the results from GPC. Two of those methods are intrinsic viscosity (IV) and melt flow rate (MFR), described below and in the next section. Neither supplies nearly as much information as is present in a GPC chromatogram, but both techniques can be easily applied in the laboratory and provide good information on the nature of the polymer produced, and the information relates to processing characteristics. Each grade of catalyst, in combination with the polymerization conditions, produces a molecular weight distribution peculiar to that catalyst. Therefore, intrinsic viscosity and MFR comparisons are best done within a set of polymers produced under similar conditions.

A polymer's IV measures its contribution to a solution's viscosity, extrapolated to conditions of infinite dilution [30]. Stated mathematically,

$$[\eta] = \lim_{\Phi \to 0} [(\eta - \eta_0) / \eta_0 \Phi]$$

where

$[\eta]$ = intrinsic viscosity, with units of deciliter/g, $dL.g^{-1}$
η = viscosity at concentration Φ, $dL.g^{-1}$
η_0 = viscosity of solvent, $dL.g^{-1}$
Φ = concentration of polymer, g/mL

The intrinsic viscosity is directly related to polymer size and to polymer processability. It is a precise and inexpensive technique for the laboratory and is determined by measuring the flow of a polymer solution using a glass capillary kinematic viscometer. This

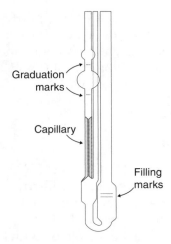

Figure 2.11 Ubbelohde viscometer. The Ubbelohde viscometer is a precision glassware tool for measuring kinematic viscosity. Reproduced by permission of IDES, Inc. Source: http://www.ides.com/property_ descriptions/templates/images/ISO307_viscometer.gif.

is a precisely fabricated glass device which permits precise measurement of flows of viscous solutions. A common type is called the Ubbelohde tube, shown in Figure 2.11.

First, one measures the flow time for a specific volume of dilute polymer solution in decalin at 135°C. This determines the relative solution viscosity versus the decalin solvent. Then the process is repeated for one or more other dilute polymer concentrations, and the results, after conversion to a quantity known as the reduced viscosity, are plotted on a graph to get a straight line plot. The y-axis intercept, which represents the reduced viscosity at "zero" concentration, is the intrinsic viscosity.

The intrinsic viscosity can be related empirically to a viscosity average molecular weight, M_v, by the following equation:

$[\eta]_{vis} = KM_v^{\alpha}$ where K and α are empirically determined constants. This is known as the Mark-Houwink equation.

At 135°C in decalin reported values are: [31]

$$K = 0.0238 \ mL/g$$
$$\alpha = 0.725$$

The constants are specific to the solvent and temperature. The viscosity average molecular weight lies between M_n, the number

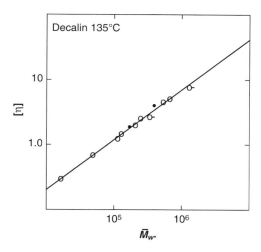

Figure 2.12 Intrinsic viscosity vs. M_w in decalin at 135°C [32]. Log of intrinsic viscosity vs. M_w for polypropylene. Open circles are isotactic polypropylene, and filled circles are atactic polypropylene. Reprinted with permission from J.B. Kinsinger, R.E. Hughes, *J. Phys. Chem.*, vol. 63, pp 2002–2007, 1959. Copyright 1959 American Chemical Society.

average molecular weight, and M_w, the weight average molecular weight, but closer to the latter. In this way, using a single piece of glassware and a heating bath, one is able to derive a polymer molecular weight from a few viscosity measurements of dilute polymer solutions. A correlation of intrinsic viscosity in decalin at 135°C vs. M_w is shown in Figure 2.12.

2.3.3.2 Measurement Method

Equipment and Chemical Requirements

- Ventilated hood
- Viscometer, Cannon Ubbelohde No. 75 [33].
- Plumb bob
- Glass oil bath with immersion heater and magnetic stir bar.
- Temperature controller
- Magnetic stir plate
- Electronic balance
- Stop watch
- Rubber suction bulb

- Raw Materials:
  - Dry polypropylene powder sample
  - Decalin stabilized with BHT

Analytical Procedure

An ASTM method exists with additional details [34]. Operations are in a ventilated hood.

1. The Ubbelohde tube must be scrupulously clean. Soak it in an oxidizing cleaning solution such as Nochromix® for several hours, rinse with distilled water, and then with acetone and air dry. Protect from dust. Use of alkali cleaning solutions is not recommended [35].
2. Heat the oil bath to 135°C +/− 1°C and maintain at that temperature with stirring.
3. Prepare 0.05, 0.1, 0.2 wt% solutions, known to 3 significant figures, of the polymer in stabilized decalin by heating at 135°C for several hours until the sample totally dissolves. Prepare the 0.2 wt% solution first, and the remainder by dilution of this solution. If any small particles remain, they must be removed by filtration or decantation.
4. Suspend the Ubbelohde tube vertically in the 135°C oil bath. It is important that the tube be vertical for accurate measurements. A simple plumb bob can assist in this. Fill the Ubbelohde tube with decalin to the filling marks in Figure 2.11. Allow the liquid temperature to equilibrate with the bath.
5. Cover the center tube with a finger and use a rubber bulb to gently suction decalin up the timing tube until past the upper graduation mark.
6. Make sure there are no bubbles, then release the suction and use a stopwatch to record the seconds it takes for the liquid meniscus to move from the upper to the lower graduation mark. Average two or more values.
7. Rinse out the tube with acetone, dry with a filtered nitrogen purge and repeat steps 4.-6. with each of the polymer solutions. The most concentrated solution should require at least twice the time of the solvent.

8. Determine the specific viscosity and reduced viscosities of each sample:
Let t = elution time sample, t_0 = elution time of solvent
Specific viscosity = η_{sp} = $(t-t_0)/t_0$
Reduced viscosity = η_{red} = η_{sp}/C where C = solution concentration in g/dL
9. Plot the reduced viscosity vs. concentration. Draw a best fit straight line through the y axis intercept (concentration = 0). This is intrinsic viscosity, η_{int} or $[\eta]$. A linear regression algorithm can be used instead of a paper plot to determine the best least squares fit for the intercept.

2.3.4   Melt Flow Rate

2.3.4.1   Introduction

In order for polypropylene to be fabricated by extrusion or injection it must be melted. The development of the melt flow rate (MFR) method provides a measure of the melt behavior with respect to flow through a capillary rheometer at 230°C under different pressures. The rheometer consists of a thermostatted capillary column with a precisely fitted piston and weights. See section 1.4 and Figure 2.13. Molten polymer is forced through a certain size orifice by the piston and the flow rate is measured.

Melt flow correlates inversely with intrinsic viscosity and molecular weight. It is largely controlled by hydrogen concentration during polymerization. See Figures 1.5 and 2.14. Viewed simply, as the polymer molecular weight increases chains become entangled, retarding polymer flow. Lower melt flow generally indicates a stronger, tougher resin, but it is more difficult to process. MFR <2 dg/min is used for pipes, MFR = 2–8 dg/min for films and fibers, and MFR> 8 for injection molding applications [36]. The most direct way to influence MFR is by the hydrogen concentration in the reactor, which acts as a chain transfer agent.

The ratio of melt flow under two different pressures or shears is known as the flow rate ratio (FRR; introduced in section 1.4). Unlike intrinsic viscosity, which is a single point method, the FRR gives some indication of how the resin will behave under differing shear conditions during processing and correlates with the polydispersity

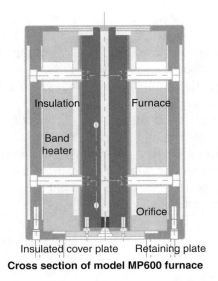

Cross section of model MP600 furnace

Figure 2.13 Crossection of a thermostatted capillary column of a melt flow plastometer. An introduction to Melt Index Testing. Molten polymer flows downward through the precisely sized orifice at the bottom of the heating zone. Reproduced by permission of Tinius Olsen.

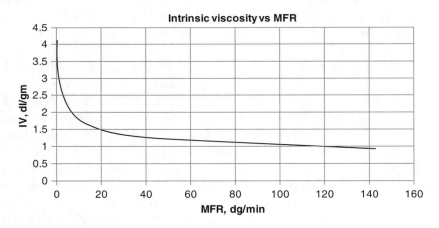

Figure 2.14 Relationship of intrinsic viscosity and MFR.

index of the polymer. Polymers with a high FRR contain short chain polymer that acts as a lubricant for the longer chains. A detailed ASTM method for MFR exists [37].

2.3.4.2   Measurement Method

Equipment and Chemical Requirements

- Extrusion Plastometer, (see photo in section 1.4)
- Temperature controller
- Timer
- Steel funnel
- Forced air oven at 45°C.
- Disposable pans
- 1% solution of Irganox® B215 in acetone (stabilizer)

Analytical Procedure

1. Place 25 g of polymer in a disposable aluminum pan and add 50mL of stabilizer.
2. Completely evaporate the solvent in a forced air oven or warm plate in the hood until a constant weight is achieved.
3. With a spatula, break the polymer mass into particles small enough to fit the extrusion chamber.
4. Set temperature of the melt indexer at 230°C and allow the indexer to reach a stable temperature.
5. Using a steel funnel pour about 5 grams of polymer into the funnel and force into the indexer with a plastic tool.
6. Remove the funnel and place the piston rod into chamber. Add a 2.16 kg weight onto the rod and let polymer begin to extrude.
7. When the piston rod marks approach the top of the chamber, cut off the extruded polymer cleanly at the bottom of the orifice. Discard this extrudate.
8. Begin timing the extrusion for 1–2 minutes.
9. Record the time exactly and cut the extrudate as before. Save this extrudate.
10. Repeat for another 1–2 minutes to collect a second sample. Calculate the average g extruded per minute for the two samples.
11. To determine the FRR, repeat the procedure with a 21.6 kg weight on the piston.
12. Calculate the melt flow rate (MFR):
    MFR = g extruded polymer/minute * 10 = g/10 minutes
13. Calculate the FRR:
    $FRR = MFR_{21.6\ kg} / MFR_{2.16\ kg}$

2.4 Polymer Bulk Density

2.4.1 Introduction

Polymer bulk density is a measure of the packing of polypropylene particles. That is, it is the weight per unit bed volume of polymer. Thus, the total volume includes void volumes within the particle and the voids between particles. Bulk density is a function of the polymer porosity, morphology and particle size distribution. These, in turn, are all a direct result of the porosity, morphology and particle size distribution of the original catalyst particles. Figure 2.15 shows a representative polymer particle, about 0.6 mm in diameter [38]. The highly textured, globule-like surface is expected to reflect the original catalyst particle from which it was produced.

The porosity of each polymer particle is also a result of the distribution of active catalyst sites within the catalyst particle. In general, high bulk density indicates high catalyst productivity. A recent study demonstrated that higher polymer bulk density derives from a more porous catalyst support which can bind active titanium throughout the particle. When titanium is bound only on the outer portions of the catalyst support, then the resulting polymer has an eggshell morphology, i.e. a partly empty interior [39]. The highest bulk density arises from a multimodal distribution of polymer spheres. The smaller spheres sit in the interstices of the packing of the larger spheres. This is shown schematically in Figure 2.16.

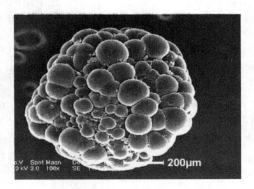

Figure 2.15 SEM micrograph of a polypropylene particle [37]. Reproduced by permission of Elsevier Ltd; Y. Chen, Y. Chen, W. Chen, D. Yang, *Polymer*, vol 47, pp. 6808–6813, 2006.

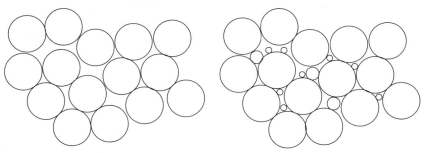

Randomly packed monodispersed spheres Randomly packed multimodal spheres

Figure 2.16 Two dimensional representation of packing of monodispersed and multimodally dispersed spheres.

Table 2.3 Mixed catalysts to improve polymer bulk density [39].

Catalyst A/B (mg/mg)	Polymer Bulk Density (g/mL)	Polymer P50 (microns)	Polymer Porosity (mL/g)
5.9/4.1	0.481	1550	0.055
10/0	0.453	2100	0.090
0/10	0.483	950	0.006

The literature describes achieving improved packing of polymer by polymerizing a mixture of two catalysts of differing particle size distribution [40]. Table 2.3 demonstrates this. Catalysts A and B were prepared from spherical supports of 21 microns and 50 micron diameters, respectively. Catalyst B is used to produce a heterophasic polymer, hence it is necessary to maintain a certain fraction of polymer voids (porosity) during the homopolymer production. By incorporating catalyst A into the homopolymerization, polymer bulk density is improved, reflecting increased plant throughput.

Higher bulk density signifies more efficient packing and is an important factor in plant design. With higher bulk density more polymer can be produced in a given polymerization reactor volume. This also means that smaller transport pipes are needed for a given throughput, and that storage silos and transport containers can be sized smaller for a given capacity. For a plant where recycling capacity of the monomer is the bottleneck, increasing polymer bulk density is a means of increasing throughput. Pelletization using screw conveyors can process more polymer per unit time, and additives blending gives more uniform distributions [41, 42]. In the extreme, low bulk density causes handling difficulties and the potential for dust explosions [43]. Commercial polypropylene has a nominal bulk density ranging from 0.35 g/mL for flakes to 0.53 g/mL for powder [44].

In certain commercial processes, highly porous, low bulk density particles are intentionally designed as part of a multistage polymerization. As described in Chapter 8, a homopolymer phase with a significant volume fraction of intraparticle voids is produced in a first stage, and then in the second stage a rubbery ethylene-propylene copolymer phase is formed in the voids in order to produce improved polymer impact and elastomeric properties. In such systems the bulk density after the first stage is lower than after the second stage.

2.4.2   Measurement Method

The bulk density measurement method is straightforward and requires only simple equipment [45, 46]. The polymer is poured via a wide mouth funnel suspended in a ring stand over a 500-mL graduated cylinder. The volume is read on the cylinder. The cylinder is then tapped with a rubber tipped implement for about a minute until no further settling is observed, and the volume is remeasured. The cylinder is then reweighed. The poured bulk density is calculated as volume/mass. A good poured bulk density is 0.45 g/mL. The difference between the poured bulk density and the tapped bulk density is a crude measure of the shape regularity of the particles; more spherical particles have a smaller difference. In principle, the highest bulk density is achieved from a bimodal or multimodal distribution of particles. The smaller particles fit into the interstices of the larger particles. See Figure 2.16.

2.5   Particle Size Distribution and Morphology

2.5.1   Introduction

The polymerization catalyst particle disintegrates during polymerization as the forces of the growing polymer fragment the catalyst. However, the polymer imprisons the catalyst fragments within itself, as shown in Figure 3.2. The net result is that one catalyst particle creates one polymer particle. So, in essence, each catalyst particle is a microreactor. Therefore, the polymer particle size distribution (PSD) is a direct image of the catalyst PSD, except magnified in size by the growth of the polymer, as shown in Figure 3.3. Much research effort has gone into designing the catalyst PSD to control the polymer PSD. A controlled polymer PSD and shape promotes high bulk density which improves equipment throughput. It reduces dusts that can foul equipment and present safety hazards. It reduces large polymer chunks that can impede handling. It improves mixing characteristics during post reactor blending with other polymers and/or additives. In the case of a spherical morphology, it provides the minimum surface/volume ratio which reduces stickiness that may arise from lower tacticity and/or incorporation of a rubbery copolymer phase in the reactor. In the extreme case of very large, uniform polymer particles, it obviates the need to pelletize the polymer at all. Spherical morphology also improves the flowability of the polymer [47]. The PSD and the morphology of the particles can provide information on the polymerization conditions within the reactor. For example, if the catalyst particle is a sphere, the presence of fractured polymer spheres may indicate that the initial phase of polymerization is too rapid, causing the catalyst particle to "explode" from insufficient heat dissipation. Fracturing can be minimized by prepolymerization.

The span of the polymer PSD is often described in terms of P95 and P5, or P90 and P10, where

> P5 = screen size through which 5 wt% passes
> P10 = screen size through which 10 wt% passes
> P50 = screen size through which 50 wt% passes
> P90 = screen size through which 90 wt% passes
> P95 = screen size through which 95 wt% passes

The ratios P90/P10 and P95/P5 give a simple way of expressing the breadth or span of the distribution:

$$\text{Span}_{95/5} = (P95 - P5)/P50 \qquad \text{Span}_{90/10} = (P90 - P10)/P50$$

A typical $\text{Span}_{90/10}$ from a 4th generation catalyst is about 0.6–0.8. The particle size distribution method described below is a straight forward classical mechanical separation of the polymer sample into different size fractions by a sieving method with an automatic shaker. See Figure 2.17. It is coupled with a visual examination

Figure 2.17 Sieve stack in an automatic shaker in an acoustically insulated cabinet. Reproduced by permission of Haver Tyler Inc. The test sieves, Ro-Tap® Shaker and cabinet are W.S. Tyler™ Equipment. Source: http://www.weavingideas.net/us/applications-products/particle-analysis-sieve-analysis/test-sieve-shakers/tyler-ro-tapr-8-rx-29.html.

of the polymer morphology. Recent instrument developments of rapid camera scanning of dry powders are also available, covering the entire particle size range of interest for polypropylene. The correlation with sieve analysis is reported to be very good [48].

2.5.2   Measurement Method

Equipment and Chemical Requirements

- Stainless steel wire-cloth sieves, 8", half-height; nominal sizes in microns: 4000, 2000, 1000, 850, 600, 425, 300, 180, 106, 75.
- Mechanical sieve shaker such as a Ro-Tap® model, with automatic timer [49] and Shaker cabinet [50] (Figure 2.17).
- Haver sample splitter (optional) [51]
- Low power stereoscope
- Brass bristle brush
- Calcium stearate powder

Analytical Procedure

1. Select a sieve stack appropriate for the sample. Tare each sieve.
2. Obtain a representative sample of about 100 g from the polymer batch. This is done by using a sample splitter or by a cone and quarter method.

3. Put the approximately 100 g of the polymer sample into a tared plastic bag, add 0.1 g calcium stearate as an antistatic agent, close the bag, and shake.
4. Pour the polymer onto the top sieve of the stack and reweigh the plastic bag to obtain the polymer total weight, W_t
5. Cover the top pan and secure the stack in the mechanical sieve shaker. Set timer for 15 minutes and close the cabinet.
6. Shake for 15 minutes.
7. Calculate the percent of total polymer on each pan and construct a curve of the weight percent through each screen. See Figure 2.18 for an example.

$$P = W_n/W_t*100$$

Where:

P = wt% polymer on n^{th} sieve
W_n = polymer weight on n^{th} sieve
W_t = total polymer weight

8. Using a low power stereoscope, examine the polymer particles from each sieve, looking for exceptional morphology, including broken particles, clumps, and excessive dust.

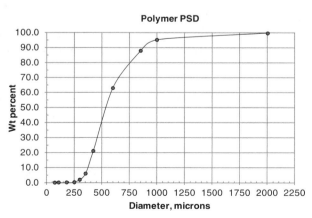

Sieve	On	Thru
microns	wt%	wt%
2000	0.2	99.8
1000	4.4	95.4
850	7.4	88.0
600	25.0	63.0
425	42.0	21.0
355	15.0	6.0
300	4.0	2.0
250	1.7	0.3
180	0.1	0.2
106	0.1	0.1
75	0.1	0.0

Figure 2.18 Example of polymer particle size distribution plot.

2.6   Questions

1. A sample of polypropylene was found to have an IV = 2.53 dL/g in decalin. What is the viscosity average molecular weight?
2. Consider a triad NMR analysis of polypropylene. Draw the triads that are possible and name them using the "m" and "r" notation. Assuming a perfectly atactic polymer sample, what is the theoretical relative intensity of the NMR resonances corresponding to the triads?
3. If an isotactic catalyst site inserts a propylene monomer correctly, then inserts a monomer with inverted stereospecificity, and then inserts the next three monomers as per the original handedness of the site, how would the sequence be expressed in "m" and "r" notation?
4. What relationship do you expect between the mmmm pentad content and the melting temperature of polypropylene, and why?
5. In Figure 2.5 there are no signals for rmmm, rrrm, mrmm, nor rrmr. Why?
6. In Figure 2.19, the chain segment has a type stereoregularity termed hemiisotacticity [52]. Propose the meaning of hemiisotactic.
7. A polypropylene sample was produced in a two hour test from a catalyst containing 2.5% titanium. The catalyst activity was 50,000 g/g. GPC determined the number average molecular weight, M_n, was 320,000. Assuming 50% of the titanium is catalytically active what was the average lifetime of a growing polymer chain? For an average chain of MW = 320,000, how fast is individual monomer insertion?
8. Why are the detectors in Figure 2.9 housed in the same cabinet as the gel permeation columns?

Figure 2.19 Polypropylene Sample.

9. A GPC chromatogram of a polypropylene sample produced with a supported catalyst produces a molecular weight curve which is deconvoluted mathematically to be the sum of three curves. Offer a hypothesis for this.

10. The theoretical density of crystalline polypropylene is 0.90 g/mL. The theoretical packing of monosized spheres is 74.05% for a hexagonal close packed (HCP) array and about 65% for randomly packed spheres. For both cases calculate the theoretical maximum bulk density of polypropylene prepared from catalyst particles that are uniformly 30 microns in diameter. Compare that value with an experimental tapped bulk density of 0.50 g/mL produced from a real-world supported catalyst. Propose at least two different reasons to account for the difference.

11. A transport container is being loaded with polypropylene powder. How might the capacity of the container be increased?

12. A sample of polypropylene powder has a poured bulk density of 0.43 g/mL. However, the portion of polymer >450 microns has a bulk density of 0.40 g/mL. What might be a possible reason?

13. Why is the PDI for a metallocene- based polypropylene more narrow than for a resin from a titanium-supported Ziegler Natta catalyst?

14. Using the PSD in Figure 2.18, calculate the Span$_{90/10}$ by linear interpolations.

15. Examination of a sample of polymer taken from a commercial production train showed the existence of unexpected clumps of polymer with evidence of fragmentation. Provide an hypothesis that might explain the clumps.

16. An investigator is interested in measuring polymer density. Which one of the techniques discussed in this chapter would be most useful in assisting that measurement?

References

1. See for example, M.P. Stevens, *Polymer Chemistry*, 3rd ed., Oxford University Press, New York, Chapter 5, 1999; A. Peacock and A. Calhoun, *Polymer Chemistry*, Chapter 5, Hanser (Munich), Chapter 5, 2006; D. Tripathi, *Practical Guide to Polypropylene*, Rapra Technology Ltd., 2002.

2. The methods can be applied to copolymers. However, microstructure determination is decidedly more complex.

3. A. Zambelli, P. Locatelli, E. Rigamonti, Macromolecules, vol 12, issue 1, pp. 156–159, 1979.

4. Regioirregular insertion is believed to result in a dormant Ti site, ultimately resulting in chain transfer/termination with hydrogen or aluminum alkyl. Thus, while regioirregular insertion is infrequent, it plays a large role in the control of the molecular weight distribution and the effectiveness of hydrogen in molecular weight control. See: F. Shimizu, J.T.M. Pater, P.M. van Swaaij, G.J. Weickert, *J. Appl. Polym. Sci.* 83, p. 2669, 2002 and references therein; J.C. Chadwick, F.P.T.J. van der Burgt, S. Rastogi, V. Busico, R. Cipullo, G. Talarico, J.J.R. Heere *Macromolecules*, 37, 9722–9727, 2004.

5. In principle the last inserted monomer, which generates a chiral teriary carbon, can control chirality in the next inserted monomer, but experimental evidence, principally from C^{13} NMR, is consistent with control at the catalytic center.

6. *Polypropylene Handbook*, 2nd edition, Nello Pasquini editor, Hanser Publishers p. 156, 2005.

7. D. Tripathi, *Practical Guide to Polypropylene,* Rapra Technology Ltd., p. 21, 2002.

8. http://winter.group.shef.ac.uk/crystal/projects-ug/sillitoe/html/history.htm; R. Freeman, *Concepts in Magnetic Resonance*, vol 11, issue 2, pp. 61–70, 1999;

9. *High Resolution NMR*, 3rd Ed., E.D. Becker, Academic Press, New York, NY, 1999; NMR *Spectroscopy: Basic Principles, Concepts, and Applications in Chemistry*, 2nd Edition, by Harald Gunther. John Wiley and Sons, 1998.

10. Recent work has shown that high field ^{13}C NMR is capable of resolving heptad and even nonad structural assemblies. See V. Busico, R. Cipullo, G. Monaco, G. Talarico, M. Vacatello, J.C. Chadwick, A.L. Segre, and O. Sudmeijer, *Macromolecules,* vol. 32, 4173–4182, 1999.

11. G. Natta, G. Mazzanti, G. Crespi, G. Moraglio, *Chim. Ind.* vol. 39, p. 531, 1957.

12. *Polypropylene Handbook*, 2nd edition, Nello Pasquini editor, Hanser Publishers 2005, pp. 361–363.

13. Early extractions also used isooctane. J. Boor Jr., *Ziegler Natta Catalysts and Polymerizations*, Academic Press, 1979, p. 43.

14. R. Paukkeri, A. Lehtinen, *Polymer*, vol. 35, number 8, pp. 1673–1679, 1994.

15. The polymer is milled by combining it with crushed dry ice, cooling, and then feeding the mixture into the mill.

16. http://technologylodging.com/wp-content/uploads/2010/05/SoxhletExtractor.gif

17. http://www.gardco.com/pages/dispersion/ml/wileyminimill.cfm

18. ASTM D5492–94. Standard Test Method for Determinatio of Xylene Solubles in Propylene Plastics.
19. Automated instrumentation: http://www.polymerchar.com/xylene_solubles
20. C. Meverden, A. Schnitgen, S. Kim, US Patent 7465776.
21. V. Virkkunen, Polymerization of Propylene with Heterogeneous Catalyst, Active Sites and Corresponding Polypropylene Structures, Academic Dissertation, U. of Helsinki, 2005.
22. S. Kim, US Patent 7056592.
23. J.C.J. Bart, *Plastics Additives Advanced Industrial Analysis*, IOS Press, 2006, pp. 704–715.
24. http://www.oxford-instruments.com
25. *Polypropylene Handbook*, 2nd edition, Nello Pasquini editor, Hanser Publishers 2005, p. 308.
26. See ASTM D 6474.
27. M. Dupire, J. Michel, US Patent 6723795.
28. For recent examples of TREF applied to polypropylene, see: I. Amer, A. van Reenen, *Macromol. Symp.* vol. 282, 33–40, 2009; P. Viville, D. Daoust, A.M. Jonas, B. Nysten, R. Legras, M. Dupire, J. Michel, G. Debras, *Polymer*, vol. 42, 1953–1967, 2001; M. Kakugo, T. Miyatake, Y. Naito, K. Mizunuma, *Macromolecules*, vol. 21, pp. 314–319, 1988.
29. A. Striegel, W.W. Yau, J.J. Kirkland, D.D. Bly *Modern Size-Exclusion Liquid Chromatography: Practice of Gel Permeation and Gel Filtration Chromatography*, 2nd Edition, John Wiley & Sons, Inc., 2009; S. Mori, H.G. Barth, *Size Exclusion Chromatography*, Springer Verlag, 1999.
30. For a discussion of intrinsic viscosity see: http://pslc.ws/macrog/vis.htm; Dabir S. Viswanath, Tushar K. Ghosh, Dasika H.L. Prasad, *Viscosity of Liquids: Theory, Estimation, Experiment, and Data*, Springer Verlag NY, 2006.
31. *Handbook of Polyolefins*, 2nd edition, C. Vasile editor, Marcel Dekker Inc., p. 366, 2000.
32. J.B. Kinsinger, R.E. Hughes, *J. Phys. Chem.*, vol. 63, pp. 2002–2007, 1959.
33. The size of the viscometer correlates with an expected viscosity range for the solution. Other size tubes are available for less viscous and more viscous solutions, respectively: 25, 50, 75, 100, 150 etc. As the number increases the interior diameter of the tube increases.
34. ASTM D446–93.
35. Cleaning & care instructions are available at http://www.cannonin-strument.com/Cleaning.htm
36. D. Tripathi, *Practical Guide to Polypropylene*, Rapra Technology Ltd., 2002, p. 76.
37. ASTM D1238–04.
38. Y. Chen, Y. Chen, W. Chen, D. Yang, *Polymer*, vol 47, pp. 6808–6813, 2006.

39. E.J. Dil, S. Pourmadian, M. Vatankhah, F.A. Taromi, *Polymer Bulletin,* 64 445–457, 2010.
40. G. Collina, O. Fusco, E.C. Carrion, A. Gil, V. Dolle, H. Klassen, K.H. Kagerbauer, US Patent 7138469.
41. G. Leicht, A. Tanchi, US Patent 3893989.
42. A.D. Holiday, US Patent 3663674.
43. J. Rosch, J. Wunsch US Patent 6096845.
44. http://www.asiinstr.com/technical/Material_Bulk_Density_Chart_P.htm
45. More sophisticated equipment are available; see http://www.labhut.com/products/powder_testers/scott.php
46. See.ASTM D-1895 for a reference method.
47. A. Sharma, S. Singh, G. Singh, V.K. Gupta, Polymer-Plastics Technology and Engineering, vol 50, pp. 418–422, 2011.
48. http://www.horiba.com/us/en/scientific/products/particle-characterization/particle-shape-analysis
49. http://www.wstyler.com/html/ro_tap.html
50. Noise from the action of the shaker is appreciable; a cabinet or remote location is desirable.
51. Available from W.S. Tyler. See http://www.weavingideas.net/us/applications-products/particle-analysis-sieve-analysis/further-laboratory-equipment/sample-splitters.html
52. J. Ewen, US Patents Nos. 6369175, 5036034.

3

Ziegler-Natta Catalysts

3.1 A Brief History of Ziegler-Natta Catalysts

As mentioned in Chapter 1, the Ziegler-Natta catalyst is so named in recognition of the pioneering work in the 1950s of Karl Ziegler in Germany and Giulio Natta in Italy. Interestingly, the origins of modern polypropylene may be traced to work by Ziegler on polyethylene catalysts in 1953. Ziegler discovered the basic catalyst systems that polymerize ethylene to linear high polymers (more below). However, an early experiment with propylene was deemed unsuccessful. Ziegler decided to concentrate on expanding knowledge of catalysts in ethylene polymerization and postponed further work on propylene. Natta was then a professor at the Institute of Industrial Chemistry at Milan Polytechnic and a consultant with the Italian company Montecatini. Natta had arranged for a cooperative research and licensing agreement between Ziegler and Montecatini. Through this arrangement, he learned of Ziegler's success with ethylene polymerization and pursued propylene polymerization aggressively. Natta succeeded in producing crystalline polypropylene and determining its crystal structure in the first half of 1954. Ziegler and Natta were jointly awarded the Nobel Prize in Chemistry in 1963 for their work in polyolefins.

Though Ziegler had an enduring interest in metal alkyl chemistry going back to the 1920s, it was not until the late 1940s that he discovered the so-called "aufbau" (growth) reaction. The aufbau reaction occurs in the essential absence of transition metal compounds and is not capable of producing high molecular weight polyethylene. However, it served as the crucial precursor discovery that ultimately made possible modern Ziegler-Natta catalysts used worldwide today for manufacture of polypropylene. In the aufbau reaction, triethylaluminum (TEAL) reacts with ethylene *via* multiple insertions to produce long chain aluminum alkyls with even numbered carbon chains. Under proper conditions, β-elimination occurs resulting in formation of α-olefins. The aluminum hydride moiety resulting from elimination is then able to add ethylene to start a new chain. If the long chain aluminum alkyls are first air-oxidized and then hydrolyzed, α-alcohols are formed. These reactions collectively are called "Ziegler chemistry" and are illustrated in Figure 3.1 [1]. Ziegler chemistry forms the basis for present-day production of many millions of pounds of α-olefins, used as comonomers for LLDPE and VLDPE, and α-alcohols, intermediates for detergents and plasticizers.

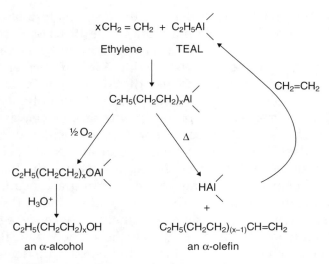

Figure 3.1 The Ziegler growth ("aufbau") reaction. (Reproduced from "*Introduction to Industrial Polyethylene*," with permission of Scrivener Publishing, LLC and John Wiley & Sons, Inc., 2010.)

Ziegler and coworkers at the Max Planck Institut für Kohlenforschung (Coal Research) in what was then Mülheim, West Germany were working to expand the scope and utility of the aufbau reaction. It was during this endeavor in 1953 that they serendipitously discovered the so-called "nickel effect." This term stemmed from the observation that nickel in combination with TEAL catalyzes dimerization of ethylene to produce 1-butene. Accounts vary on the source of nickel in the formative experiments. It was eventually attributed to trace nickel extracted from the surface of the stainless steel reactor in which early reactions were conducted.

Ziegler then launched a systematic study of the influence of transition metal compounds combined with aluminum alkyls on reactions of ethylene. Heinz Breil, a young graduate student, was assigned responsibility. Though early experiments were disappointing, Breil persisted and eventually combined zirconium acetylacetonate with TEAL and produced linear polyethylene as a white powder. Later, Heinz Martin combined titanium tetrachloride with TEAL and obtained a highly active catalyst for ethylene polymerization. (Indeed, in the initial experiment, the catalyst was so "hot" that the polymer was charred.) It was abundantly clear from early work of Ziegler's group that the linear polyethylene produced with their catalysts exhibited vastly different properties relative to the highly branched polyethylene produced with free radical initiators. It was not until June of 1954 that Ziegler finally returned to propylene, but by then Natta had already succeeded and filed for patents [2]. (This turn of events resulted in hard feelings and ended the friendship between Ziegler and Natta. See Chapter 8 of reference 2.)

Though combinations of titanium salts and aluminum alkyls have been modified and refined over the past five plus decades, it is remarkable that such combinations capture the essential character of most early 21st century Ziegler-Natta catalysts for polypropylene. The genesis of Ziegler-Natta polyolefin catalysts has been authoritatively described by McMillan [2], Seymour [3, 4], Boor [5] and Vandenberg and Repka [6].

3.2   Definitions and Nomenclature

Ziegler-Natta catalysts are broadly defined as combinations of a transition metal compound from Groups 3–12 of the Periodic Table with an organometallic compound from Groups 1, 2 or 13.

These binary catalyst systems are not merely "complexes." Rather, substantial reactions occur between the organometallic and transition metal compounds. Of course, not all combinations work as polymerization catalysts, but patent claims were made very broad to maximize coverage. Most commercial Ziegler-Natta catalysts are heterogeneous solids, but some (primarily those derived from vanadium compounds) are homogeneous (soluble in hydrocarbons). After polymerization, the catalyst is interspersed throughout the polymer and cannot be isolated. Hence, recycle of Ziegler-Natta catalysts is not practical. In reduced form, the transition metal component is called the "catalyst," while the organometallic part is called the "cocatalyst," or less commonly, the "activator." In most cases, a titanium compound (frequently titanium tetrachloride, $TiCl_4$) and an aluminum alkyl are used. Alone, each component is incapable of converting olefins to high polymers. Interactions between catalyst and cocatalyst will be addressed in the discussion of the mechanism of Ziegler-Natta polymerization in section 3.7 and in Chapter 5.

Some catalyst chemists consider single site catalysts to be a subset of Ziegler-Natta catalysts, because most involve combinations of transition metal compounds with Group 13 organometallics. However, most single site catalysts are homogeneous and produce polypropylene with properties that are quite distinct from polypropylene made with Ziegler-Natta catalysts. Additionally, single site catalysts are believed to exist as cations during the polymerization of olefins, unlike Ziegler-Natta catalysts. Accordingly, single site catalysts are discussed separately (see Chapter 6).

Catalyst efficiency is expressed in different ways. It is variously called "activity," "yield," "productivity" and "mileage," but there are subtle differences. Activity is sometimes expressed as weight (g, kg or lb) of polypropylene per wt of catalyst per unit time. (Average catalyst residence times in commercial polypropylene reactors can range from a few minutes to several hours, depending on process.) Productivity, mileage and yield usually mean the weight of polymer produced per weight of catalyst, regardless of time. The latter method is frequently used for expressing efficiency of industrial catalysts. Another way of indicating catalyst activity is the unit named in honor of Karl Ziegler. In the parlance of polypropylene, the "Ziegler" is usually expressed as g PP per millimole of transition metal (often Ti) per hour per atmosphere of propylene written as:

$$\text{g PP/mmole Ti-h-bar } C_3H_6$$

The Ziegler is used in journal literature and academia.

A few additional comments are warranted on nomenclature. Since discovery in the 1950s, Ziegler-Natta catalysts have been known by a variety names such as "coordination catalysts," "coordinated anionic catalysts" (originally proposed by Natta), "Ziegler catalysts" (applied mostly to polyethylene) and "Natta catalysts" (applied mostly to polypropylene). The term "Ziegler chemistry" has also been applied to reactions of aluminum alkyls with olefins, especially the aufbau reaction [1]. In this text, we will use the notation "Ziegler-Natta catalysts" to encompass propylene polymerizations employing heterogeneous catalysts produced by combining transition metal salts with metal alkyls. As previously noted, relatively few Ziegler-Natta catalysts are homogeneous. (For reasons previously mentioned, single site catalysts will be considered separately).

3.3   Characteristics of Ziegler-Natta Catalysts

Ziegler-Natta catalysts are not pure compounds. Most are heterogeneous inorganic solids, essentially insoluble in hydrocarbons and other common organic solvents, making them difficult to study. In reduced form, Ziegler-Natta catalysts are typically highly colored (from violet to gray to brown) powdery or granular solids. Many produce smoke upon exposure to air and may be violently reactive with water. Because Ziegler-Natta catalysts may be rendered inactive ("poisoned") even by traces of oxygen and water, they must be handled under an inert atmosphere (usually nitrogen). Consequently, polyolefin manufacturers typically combine components on site to produce the active Ziegler-Natta catalyst system. Polypropylene manufacturers either produce their own catalyst on-site or purchase catalyst from a licensor. Cocatalyst is most commonly purchased from metal alkyl suppliers. See Chapter 5. In most cases, Ziegler-Natta catalyst systems are not storage stable after the two components are combined. Catalyst and cocatalyst are usually combined in the polymerization reactor. Catalyst recipes are highly proprietary and it is often difficult to discern actual practices from what McMillan called the "fairyland of patent applications" [7].

An important characteristic of heterogeneous polypropylene catalysts is the phenomenon of particle replication. Particle size distribution (psd) and morphology of the catalyst are reproduced in the polymer. If the catalyst is finely divided, the polymer will also be fine and may cause handling problems. If the catalyst contains agglomerates of oversized particles, so too will the polymer. Morphology replication is illustrated in Figure 3.2. As the polymer accumulates on the surface of the catalyst particle and within its pores, the growing polypropylene forces the catalyst to disintegrate into small fragments. These fragments remain entrained in the growing polymer particle, and the fragmentation exposes new active catalyst surface, accelerating the polymerization process.

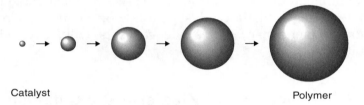

Catalyst   Polymer

Figure 3.2 The phenomenon of replication. As the polymer particle grows, it assumes the morphology of the catalyst particle. This is referred to as "replication," i.e., a spherical catalyst results in a spherical polymer particle.

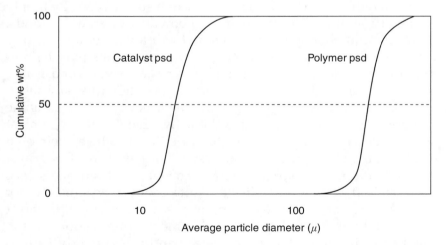

Figure 3.3 Replication of particle size distribution. Particle size distribution of polymer mirrors that of catalyst particles. In this example, a catalyst with ~ 40 μ average diameter grows to a polymer particle of ~500 μ. (Reproduced from *Introduction to Industrial Polyethylene*, with permission of Scrivener Publishing LLC).

The ultimate result is that the shape of the catalyst particle is replicated by the polymer particle on a larger scale, with the catalyst particle dispersed throughout the polymer particle. Thus, ideally each catalyst particle produces one, much larger, polymer particle. Figure 3.3 shows how psd of the catalyst is mirrored in the polymer.

Ziegler-Natta catalysts are capable of polymerizing olefins under very mild conditions. Consider that conditions for free radical polymerization of ethylene often require temperatures >200°C and pressures of >140 MPa (~20,000 psig). In contrast, Ziegler's group showed that Ziegler-Natta catalysts are capable of polymerizing ethylene at atmospheric pressure and ambient temperature [8]. Propylene polymerizations may also be conducted under relatively mild conditions; for example, propylene polymerizations are often conducted at ~70°C and pressures below 500 psig.

3.4   Early Commercial Ziegler-Natta Catalysts

Titanium tetrachloride, a clear colorless liquid under ambient conditions, was the logical choice as the raw material for early Ziegler-Natta catalysts. $TiCl_4$ (aka "tickle 4") was (and still is) manufactured in enormous volumes as a precursor to titanium dioxide used as a pigment for paint. Consequently, $TiCl_4$ was readily available and relatively inexpensive. Moreover, $TiCl_4$ had been shown by Ziegler and coworkers to produce some of the most active catalysts. Though sometimes called a catalyst, it is more accurate to call $TiCl_4$ a "precatalyst," since it must be reduced and combined with a cocatalyst to become active.

Early (1960–1965) commercial Ziegler-Natta catalysts ("first generation") were produced by reduction of $TiCl_4$ using metallic aluminum, hydrogen or an aluminum alkyl. Simplified overall equations are listed below:

$$TiCl_4 + \tfrac{1}{3}\,Al \rightarrow TiCl_3 \bullet \tfrac{1}{3}\,AlCl_3 \text{ [also written } (TiCl_3)_3 \bullet AlCl_3] \qquad (3.1)$$

$$2\,TiCl_4 + H_2 \rightarrow 2\,TiCl_3 + 2\,HCl \qquad (3.2)$$

$$2\,TiCl_4 + 2\,(C_2H_5)_3Al_2Cl_3 \rightarrow 2\,TiCl_3 \downarrow + 4\,C_2H_5AlCl_2 + C_2H_4 + C_2H_6 \qquad (3.3)$$

The predominant product in each case was titanium trichloride (aka "tickle 3"), an active catalyst for olefin polymerization. However, even $TiCl_3$ required a cocatalyst (or "activator") to become an effective catalyst for polymerization of propylene. The preferred cocatalyst was diethylaluminum chloride (DEAC). $TiCl_3$ from eq 1 contains co-crystallized aluminum trichloride. $TiCl_3$ from eq (3.3) may also contain small amounts of complexed aluminum alkyl. Products from eq (3.1) and (3.2) were supplied commercially by companies such as Stauffer Chemical and Dart (both now defunct). Catalyst from eq (3.3) was manufactured on site by a major polypropylene producer of the time.

Activity was poor (500–1000 g polypropylene/g catalyst) for these antiquated catalysts. Early manufacturers of polypropylene were required to post-treat product to remove atactic polymer and residues containing transition metals and acidic chloride compounds. The latter residues could cause discoloration of polymer or corrosion of downstream processing equipment. In the mid-1960s, ball-milling of catalysts was introduced and resulted in increased surface area and marginally higher activity. Catalysts from eq (3.1)–(3.3) are now largely obsolete. However, $TiCl_3 \cdot \frac{1}{3}$ $AlCl_3$ from eq (3.1) is still supplied commercially, but at greatly reduced volumes.

$TiCl_3$ is a highly colored solid that exists in several crystalline forms, designated as alpha (α), beta (β), gamma (γ) and delta (δ). The α, γ and δ forms have layered crystal structures and are violet. The β form has a linear structure and is brown. Crystal structures of these compositions have been reviewed [9, 10].

An intermediate improvement in PP catalysts ("second generation") was made in the early 1970s. These catalysts were originally developed by Solvay (see p 227 of reference 16 and p 41 of reference 11) [11]. They are called "Solvay catalysts" or "precipitated catalysts" and typically employ DEAC as cocatalyst. Relative to early Ziegler-Natta catalysts, such catalysts showed high surface area, spherical morphology, greater stereochemical control and improved activity. However, because of the way they were produced, early versions were very fragile. Prepolymerization (see section 3.6) improved the "robustness" of the catalyst. Modern versions are still in use today under the trade name "Lynx" catalysts supplied commercially by BASF. ("Lynx" is the trade name used for a broad range of polypropylene catalysts, some of which are supported.)

3.5   Supported Ziegler-Natta Catalysts

In non-supported catalysts, most of the transition metal (usually titanium) does not become active. It has been shown that $\geq 90\%$ of the transition metal sites are inactive (see Chapter 4). Consequently, non-supported catalysts show dramatically lower average activity per transition metal. This could be caused by the transition metal not being in a correct chemical environment or because of inaccessibility to active centers.

A major improvement occurred in the 1970s when supported Ziegler-Natta catalysts began to emerge. Leading polyolefin producers of the time (Solvay & Cie, Hoechst, BASF, Mitsui, Montecatini Edison, Shell, etc.) developed many of these catalysts [12]. Of course, several have morphed into present-day companies, such as LyondellBasell and INEOS. The early supported Ziegler-Natta catalysts are now commonly designated "third generation" catalysts. Further improvements have been made in supported catalysts allowing even higher efficiency, better control of morphology and particle size distribution, and excellent stereochemical control. These are designated fourth and fifth generation catalysts. Generations of polypropylene catalysts will be discussed in greater detail in Chapter 4.

Many inorganic compounds were tested as supports for polyolefin catalysts, but magnesium salts provided the most serviceable catalysts for polypropylene. Magnesium compounds such as MgO, $Mg(OH)_2$, $HOMgCl$, $ClMgOR$ and $Mg(OR)_2$, have been used, but anhydrous $MgCl_2$ is the most widely employed in industrial Ziegler-Natta catalysts. Chien described $MgCl_2$ as the "ideal support" for titanium active centers and summarized several reasons for the desirability of $MgCl_2$ as a support [13]. For example, Chien cited the similarity of crystal structures between $MgCl_2$ and $TiCl_3$. Titanium chloride active centers are chemisorbed on the surface of the magnesium compound. Worldwide, $MgCl_2$-supported Ziegler-Natta catalysts have become the most important catalysts for polymerization of propylene.

Supported catalysts result in dispersed active centers that are highly accessible. Catalyst activity is greatly increased (>10,000 g polypropylene/g catalyst). TEAL is the preferred cocatalyst for supported Ziegler-Natta catalysts. Transition metal residues in polypropylene produced with modern supported catalysts are very low (typically <5 ppm), obviating post-reactor treatment of polymer.

As mentioned above, supports make it possible to control psd and morphology of the catalyst, which is reflected in the polymer, previously illustrated in Figures 3.2 and 3.3. Proper control of catalyst psd is desirable because it translates into narrower polymer psd, thereby minimizing large (and small) particles. Polymer transfer in commercial operations is achieved primarily by pneumatic conveyance. Hence, large particles can lead to flow problems, such as plugged transfer lines. Fine particulates may also cause problems, such as clogged filters and heightened possibility of dust explosions. Morphology control increases bulk density of the finished resin and is believed to improve fluidization dynamics in gas phase processes.

3.6   Prepolymerized Ziegler-Natta Catalysts

A technique called "prepolymerization" is practiced with selected psd controlled catalysts used in slurry and gas phase processes. The catalyst is suspended in propylene or a suitable solvent (usually a C_3–C_7 alkane) and exposed to cocatalyst, monomer, and, optionally, comonomer and hydrogen under mild (low T and P) conditions in a separate, smaller reactor [14]. Heat removal is very efficient. Prepolymerization is typically allowed to proceed until the original catalyst comprises up to about 30% of the total weight of the composition. Many prepolymerized catalysts have limited storage stability and are ordinarily introduced without delay to the large-scale reactor. Prepolymerization provides several advantages:

- Preserves catalyst morphology by holding the catalyst particle (and later the polymer particle) together, making the catalyst/polymer particle more "robust," *i.e.*, less prone to fragmentation (which may produce undesirable fines)
- Increases bulk density
- Provides a heat-sink during the early stages of polymerization so that the catalyst does not overheat in a gas phase reaction.

To make the polymer particle more robust, the prepolymer is most effective when it is of high molecular weight, relative to the main polymer production.

3.7 Mechanism of Ziegler-Natta Polymerization

Despite passage of more than half a century since the basic discoveries, the mechanism of Ziegler-Natta polymerization is still not fully understood. As in all chain-growth polymerizations, the basic steps are initiation, propagation and termination (chain transfer). Presented below is a brief survey of the principal features of Ziegler-Natta polymerization of propylene. The reader is directed to excellent discussions of the mechanism of propylene polymerization in the literature for more detailed information. (See Boor [9], Collman, *et al.* [14,15] and Krentsel, *et al.* [15, 16] for historical views, and references [16–18] for more recent discussions).

Cossee and Arlman were among the first to propose a comprehensive mechanism for Ziegler-Natta catalysis and supported their proposals with molecular orbital calculations [19, 20]. The Cossee-Arlman proposal involves a "migratory alkyl transfer" and, with some refinements, remains the most widely cited mechanism for Ziegler-Natta catalysis [21, 22].

The reduced form of titanium is octahedral and contains open coordination sites (□) and chloride ligands on crystallite edges. Initiation begins by formation of an active center, believed to be a titanium alkyl. Alkylation by TEAL cocatalyst produces an active center:

$$(C_2H_5)_3Al \quad + \quad \overset{Cl}{\underset{\diagup}{\mid}} Ti \diagdown \square \quad \longrightarrow \quad \overset{\overset{\delta^+ \diagup C_2H_5}{Cl \cdots Al \diagdown C_2H_5}}{\underset{\diagup}{\mid}} Ti \cdots C_2H_5$$

$$\downarrow$$

$$\overset{\square}{\underset{\diagup}{\mid}} Ti \diagup - C_2H_5 \quad + \quad (C_2H_5)_2AlCl \tag{3.4}$$

The alkyl migrates (rearranges) such that an open coordination site moves to a crystallite edge position [21, 22]. Coordination of propylene monomer occurs to create a π-complex as in eq 3.5.

Subsequent addition across propylene results in the propagating species:

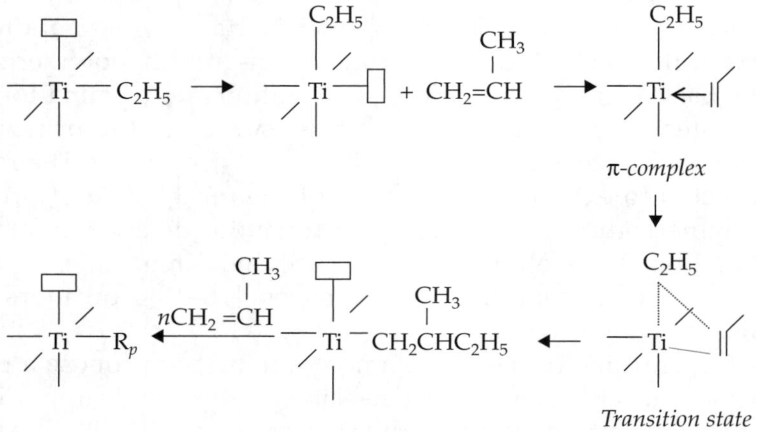

$$R_p = \sim(CH_2CH)_{n+1}C_2H_5, \text{ a polymer chain} \qquad (3.5)$$

with CH_3 substituent shown on the chain.

Schematics depicting regiochemistry and stereochemistry are shown in Figures 3.4 and 3.5.

Because titanium-carbon σ-bonds are known to be unstable, a different mechanism that invokes coordination of the aluminum alkyl to the titanium alkyl has been postulated. It is suggested that the titanium alkyl is stabilized by association with the aluminum alkyl. Coordination also accommodates the well-known propensity

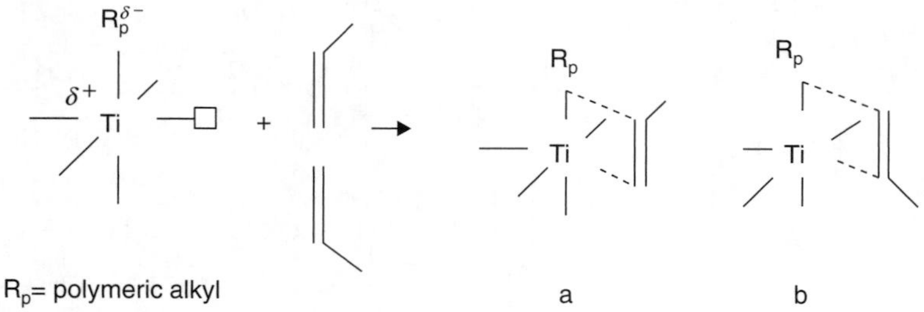

Figure 3.4 Regioselectivity in propylene polymerization. Schematic depiction of 2 possible modes of addition of Ti-C bond in the propagation step of ZN polymerization of propylene. Complex (a) is the predominant mode ("head-to-tail" addition).

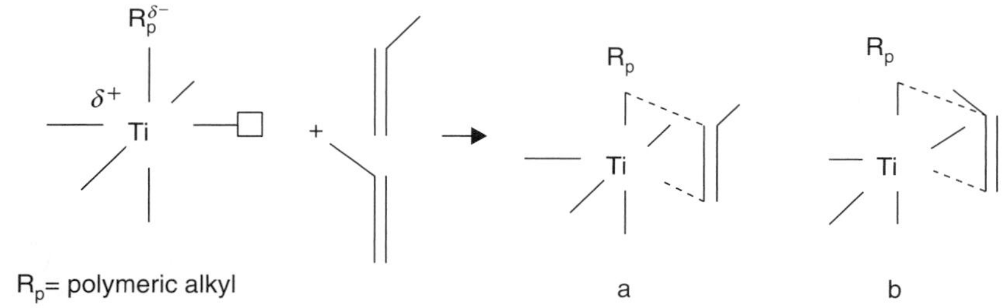

Figure 3.5 Stereoselectivity in propylene polymerization. Simplified representation of "head-to-tail" addition of Ti-C across propylene with 2 possible methyl group orientations. Stereochemistry of the methyl group determines whether PP is isotactic, atactic, syndiotactic, etc.

of aluminum alkyls to associate [23]. This is known as the "bimetallic mechanism," and essential features were originally proposed by Natta and other workers in the early 1960s [24]. Basic steps are similar to the Cossee-Arlman mechanism. The principal difference is participation of the aluminum alkyl. However, polymerization is still believed to occur by insertion of C_3H_6 into the Ti-C bond (rather than the Al-C bond). Key steps are illustrated in eq 3.6 below:

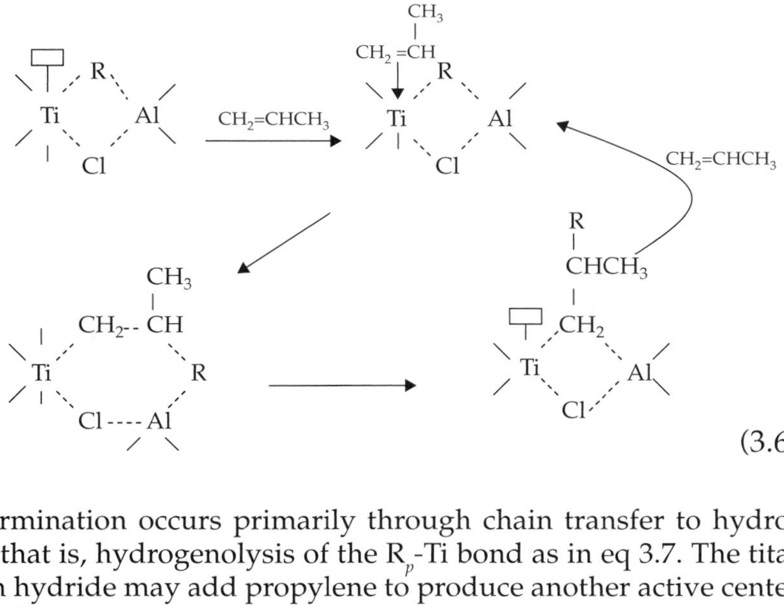

$$(3.6)$$

Termination occurs primarily through chain transfer to hydrogen, that is, hydrogenolysis of the R_p-Ti bond as in eq 3.7. The titanium hydride may add propylene to produce another active center for polymerization.

$$\overset{|}{\underset{|}{Ti}} - R_p \ + \ H_2 \ \longrightarrow \ \overset{|}{\underset{|}{Ti}} - H + R_pH$$

$$\downarrow CH_2 = CHCH_3 \qquad (3.7)$$

$$\overset{|}{\underset{|}{Ti}} - CH_2CH_2 \overset{CH_3}{|}$$

Chain termination may also occur by β-elimination with hydride transfer to titanium (eq 3.8), by β-elimination with hydride transfer to monomer (eq 3.9) and chain transfer to aluminum alkyl (eq 3.10).

$$\overset{|}{\underset{|}{Ti}} \longrightarrow \overset{|}{\underset{|}{Ti}} - H \ + \ CH_2{=}CR_p \overset{CH_3}{|} \qquad (3.8)$$

$$\overset{|}{\underset{|}{Ti}} \longrightarrow \overset{|}{\underset{|}{Ti}} - CH_2CH_2 \overset{CH_3}{|} \ + \ CH_2{=}CR_p \overset{CH_3}{|} \qquad (3.9)$$

$$(C_2H_5)_3Al + \overset{|}{\underset{|}{Ti}} - R_p \longrightarrow \overset{|}{\underset{|}{Ti}} \cdots R_p \longrightarrow \overset{|}{\underset{|}{Ti}} \ + \ (C_2H_5)_2AlR_p$$

$$(3.10)$$

The aluminum alkyl product from eq (3.10) containing the polymeric chain (R_p) will undergo hydrolysis or oxidation/hydrolysis when the resin is exposed to ambient air, analogous to the aufbau chemistry depicted in Figure 3.1. This chemistry results in polymer molecules with methyl and ~CH_2OH end groups, respectively. However, concentrations are miniscule, because the vast majority of chain termination occurs by eq (3.7)–(3.9).

In chain transfer/termination reactions illustrated in eq (3.7)–(3.10), the component containing the transition metal is still an active catalyst. Thus, each active center may produce hundreds or thousands of polymer chains.

Ziegler-Natta catalysts are the most important transition metal catalysts for production of polypropylene. Indeed, at this writing, it would not be practicable to manufacture stereoregular polypropylene needed for the global market without Ziegler-Natta catalysts. This may change as single site catalyst technology continues to evolve and mature, but Ziegler-Natta catalysts will remain essential to polypropylene manufacture well into the 21st century.

3.8   Questions and Exercises

1. How did Karl Ziegler's research on the "aufbau" reaction contribute to the development of modern polypropylene catalysts?
2. Why is titanium tetrachloride the dominant choice for raw material for industrial polypropylene catalysts?
3. What are the advantages of "prepolymerization"?
4. What is the main difference between the Cossee-Arlman and the "bimetallic" mechanisms of ZN polymerization of propylene?
5. Illustrate the termination reactions of ZN polymerization of propylene. Which is the primary method of MW control for ZN polymerization of propylene?

References

1. JR Zietz, Jr., GC Robinson and KL Lindsay, *Comprehensive Organometallic Chemistry*, vol. 7, 368, 1982.
2. FM McMillan, *The Chain Straighteners*, MacMillan Press, London, 1979.

3. RB Seymour and T Cheng, *History of Polyolefins*, D. Reidel Publishing Co., Dordrecht, Holland, 1986.

4. RB Seymour and T Cheng (editors), *Advances in Polyolefins*, Plenum Press, New York, 1987.

5. J Boor, Jr., *Ziegler-Natta Catalysts and Polymerizations*, Academic Press, Inc., 1979.

6. EJ Vandenberg and BC Repka, *High Polymers*, (ed. CE Schildknecht and I Skeist), John Wiley & Sons, *29*, p. 337, 1977.

7. FM McMillan, *History of Polyolefins*, (RB Seymour and T Cheng, editors), D. Reidel Publishing Co., Dordrecht, Holland, xi, 1986.

8. FM McMillan, *The Chain Straighteners*, MacMillan Press, London, 67, 1979.

9. J Boor, Jr., *Ziegler-Natta Catalysts and Polymerizations*, Academic Press, Inc., 92, 1979.

10. N. Pasquini (editor), *Polypropylene Handbook*, Hanser, 18, 2005.

11. YV Kissin, *Alkene Polymerizations with Transition Metal Catalysts*, Elsevier, The Netherlands, 6, 2008.

12. FJ Karol, *Encyclopedia of Polymer Science and Technology*, Supp vol. 1, 120 (1976).

13. JCW Chien, *Advances in Polyolefins*, Plenum Press, New York, 256, 1987.

14. J. Kivela, H. Grande and T Korvenoja, *Handbook of Petrochemicals Production Processes*, McGraw-Hill, *16*, 45, 2005.

15. JP Collman, LS Hegebus, JR Norton and RG Finke, *Principles and Applications of Organotransition Metal Chemistry*, University Science Books, Sausalito, CA, 577, 1987.

16. BA Krentsel, YV Kissin, VJ Kleiner and LL Stotskaya, *Polymers and Copolymers of Higher α-Olefins*, Hanser/Gardner Publications, Inc., Cincinnati, OH, 6, 1997.

17. G. Cecchin, G. Morini and F. Piemontesi, *Kirk-Othmer Encyclopedia of Chemical Technology*, Wiley Interscience, New York, 26, 502, 2007.

18. E. Albizzati, G. Cecchin, JC Chadwick, G. Collina, U. Giannini, G. Morini and L. Noristi, *Polypropylene Handbook*, (N. Pasquini, editor), Hanser, 49, 2005.

19. P Cossee, *J. Catal.* 1964, *3*, 80.

20. EJ Arlman and P Cossee, *J Catal.* 1964, *3*, 99.

21. FA Cotton, G Wilkinson, CA Murillo and M Bochmann, *Advanced Inorganic Chemistry*, 6th ed., John Wiley & Sons, New York, p. 1270, 1999.

22. MP Stevens, *Polymer Chemistry*, 3rd ed., Oxford University Press, New York, 238, 1999.

23. T Mole and EA Jeffery, *Organoaluminium Compounds*, Elsevier Publishing Co., Amsterdam, 95, 1972.

24. J Boor, Jr., *Ziegler-Natta Catalysts and Polymerizations*, Academic Press, Inc., 334, 1979.

4

Propylene Polymerization Catalysts

4.1 Introduction

This chapter reviews the technological development of Ziegler-Natta catalysts beyond the introductory material in Chapters 1 and 3. For additional details the reader is referred to other texts [1–5]. The evolution of Ziegler-Natta propylene polymerization catalysts can be divided reasonably well in terms of catalyst generations. As a practical matter, the overwhelming majority of industrial production now relies upon fourth generation and later catalysts. Nevertheless, a discussion of the earlier generations provides essential context for the descriptions of the later generations.

Following the 1953 discovery by Karl Ziegler at the Max Planck Institute that ethylene polymerized in the presence of certain transition metal catalysts, work was initiated by Giulio Natta at Milan Polytechnic University, who consulted with Montecatini, to extend Ziegler's findings to other polymers, and resulted in his synthesis of crystalline polypropylene in 1954. However, there was also contemporary work ongoing at other laboratories around the world. So, from a legal perspective, two laboratories, those of Giulio Natta at Montecatini Edison and of Hogan and Banks at Phillips Petroleum, would figure most prominently in the ultimate US patent claims to the first preparation of crystalline polypropylene, a saga which required decades of legal examination to settle [2]. See also discussion in section 1.1.

Irrespective of the legal situation, it was recognized by all researchers that the original propylene polymerization catalyst systems had two very significant limitations with respect to commercialization: the catalyst polymerization activity was far too low, and the polypropylene isotacticity was too low. The practical implications were that a commercial process had to be designed both to neutralize and remove the catalyst residues, and to separate the large amount of atactic polymer from the isotactic polymer. These were substantial requirements, and the early process designs were accordingly complicated.

Thus, the essential elements in the history of polypropylene catalyst development are the searches for higher catalyst activity and for higher catalyst stereospecificity, with the added goal of controlling polymer morphology. As polypropylene began to be commercially produced, this search intensified. The research was quite costly, and the real value of a kilogram of catalyst resided, not in the commercial price of the catalyst, but in the thousands of kilograms of polymer it produced. Accordingly, the bulk of research was funded by companies interested in polypropylene production, and, although polypropylene was initially pioneered at an academic institution, most significant advances in ZN catalyst know-how from the 1950s until the present have been the product of industrial research.

As will be described in this chapter, succeeding generations of catalyst are delineated principally by advances in activity and stereospecificity, as well as improvements in polymer morphology. These catalysts have been exceedingly successful; in 2012 it is projected global demand for polypropylene will reach about 55 million mtons/yr [3]. Polymerization catalyst developments, which have occurred over the last half century, have largely been the result of Edisonian investigation, namely exhaustive trial and error. This is because Ziegler-Natta catalysts are heterogeneous or surface active materials and difficult to structurally define. As with any surface catalyzed phenomenon, there are many types of catalytic structures present, *i.e.* there is a cocktail of catalyst sites. Each type of site has its own characteristics and contributes to the overall observed macroscopic behavior. Computational and physical modeling of such complicated systems, while elegant, proved less effective than

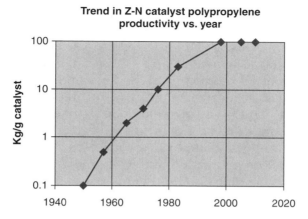

Figure 4.1 Historical catalyst productivity.

synthetic screening programs [4]. These results have been quite impressive, and today the problems of catalyst activity and stereo-specificity are essentially solved. Figure 4.1 shows an historical plot of the progress achieved in ZN catalyst activity.

Concurrent with advances in catalyst activity and stereospeci-ficity, as the commercial importance of polypropylene grew there have also been developments in tailoring the resin. This includes control of the molecular weight distribution, polymer morphology, and the incorporation of comonomers. Although the design of the polymerization process can impact these properties, it is mainly through improvements in the catalyst design that major advances are possible [3]. As new catalysts were developed, this enabled the development of improved and more sophisticated industrial pro-cesses. Production economics also improved, and it became fea-sible to design larger polypropylene plants. These factors have all contributed to the growth of world demand for this strong, light-weight, and versatile resin.

4.2   Zero Generation Ziegler-Natta Catalysts

Natta's synthesis of polypropylene in 1954 produced material with less than 50% isotactic content, using a catalytic combination of titanium tetrachloride and aluminum alkyl. For his work on poly-propylene and for Ziegler's pioneering work in polyethylene, Natta and Ziegler were awarded the Nobel Prize in 1963. Screening work on catalyst families quickly identified titanium trichloride in combination with alkylaluminum compounds as the most promising catalyst system. This is demonstrated by a selection of screening data from Natta, as shown in Table 4.1. Within three years, in 1957, the first commercial production at Montecatini had begun.

It is important to recognize the good fortune that titanium and aluminum formed the key metallic elements of the successful catalyst system. Both elements are inexpensive, readily available, non-toxic and form inert, white oxides. Had Nature decreed that, for example, tantalum and gadolinium were the key metals required, the polypropylene industry would not exist as it does today [6, 7].

Early patents focused on generation of the catalyst in the polymerization reactor. For example, US 3014018 reports: " a solution of 2.7 g of triethylaluminum ….in 100 cc heptanes and successively a solution of 1.73 g titanium tetrachloride in 200 cc heptanes are introduced into the autoclave within about 10 seconds. The autoclave is then shaken [and] propylene quickly introduced" [8]. These "zero generation" catalysts consisted largely of $TiCl_3$ in the beta crystalline phase. They had very low activity, about 20 g polypropylene/g catalyst.hr, and produced a large fraction of atactic polymer. In the same time frame others reported using catalyst slurries prepared outside the polymerization reactor that gave improved polymer crystallinity [9]. A milestone in the very early catalyst development was the discovery of molecular hydrogen as an efficient molecular weight control agent, originally identified for polyethylene [10].

Table 4.1 Isotacticity from early catalyst screening [5].

Catalyst	Metal Alkyl Cocatalyst	Polypropylene Isotactic Index
β-$TiCl_3$	TEAL	40–50
δ-$TiCl_3$	TEAL or DEAC	96–98
δ-$TiCl_3$	Diethylberylium	94–96
δ-$TiCl_3$	Diethylmagnesium	78–85
δ-$TiCl_3$	Diethylzinc	30–40
VCl_3	TEAL	73
$TiCl_2$	TEAL	75

This was essential to the development of commercial polypropylene grades.

These early catalysts have been assigned the term "zero generation" because they were preliminary versions essentially from discovery experiments. Further developments built progressively upon these "zero generation" catalysts to begin the long road toward the optimized versions that are used today to produce megaton quantities of polypropylene.

4.3 First Generation ZN Catalysts

In-situ catalyst preparation was followed quickly by the preparation of titanium trichloride slurries external to the reactor. Reduction of neat $TiCl_4$ with hydrogen at ~700°C or with aluminum metal produced the alpha (α) crystalline form. Aluminum metal reduction proved more effective than hydrogen reduction, giving a solid solution of aluminum trichloride dispersed in a titanium trichloride matrix:

$$TiCl_4 + Al(s) \rightarrow TiCl_3.1/3AlCl_3 \qquad (4.1)$$

Reduction with aluminum alkyls gave the beta form, and heating the beta form converted it into the more active gamma form. Mechanical milling or grinding of either the alpha or gamma forms produced the most active active delta form of the crystallites [24]. The interconversion of the various forms is summarized in Figure 4.2 [11].

Catalyst preparation was optimized by combining reduction with aluminum with an intensive mechanical ball milling step in

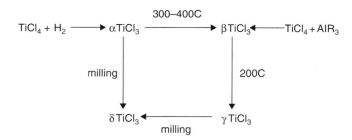

Figure 4.2 Transformations of titanium trichloride. Adapted from J. Boor Jr., *Ziegler Natta Catalysts and Polymerizations*, Academic Press, page 95, 1979.

the absence of any solvent. This increased surface area and converted the crystal structure to the delta phase. The latter change was principally responsible for raising catalyst activity about ten fold. Activation depended upon the impact energy and temperature of the milling. One early patent describes the process: "A TiCl$_3$–0.33 AlCl$_3$ catalyst components was prepared by the reduction of TiCl$_4$ with the stoichiometric amount of aluminum powder at 230°C in a steel bomb. This catalyst component was dry ball-milled with... steel balls...for 4 days" [12]. This product was demonstrated to give about 200 g polypropylene/g catalyst with an isotactic index as high as 96%. This catalyst type best delineates the first generation. A representative chart is shown in Figure 4.3, demonstrating the benefits of milling and the superiority of TiCl$_3$.1/3AlCl$_3$ vs TiCl$_3$.

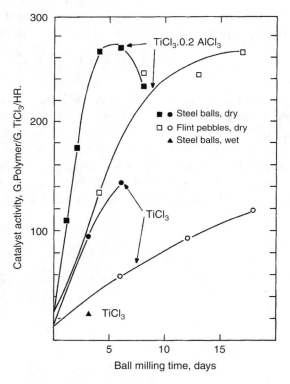

Figure 4.3 Effect of milling upon TiCl$_3$ activity [12]. Milling is much more effective for activating TiCl$_3$ reduced with aluminum, then for hydrogen reduced TiCl$_3$. Source: E. Tornqvist, A. W. Langer Jr. US Patent No. 3032510.

The nature of 1st generation catalyst synthesis did not read-ily lend itself toward incorporation into the operations of a poly-mer plant, but the improved activity and stereospecificity of $TiCl_3.1/3AlCl_3$, were significant advances. So, catalyst synthesis became a specialized manufacturing operation, separate from the polymerization process. This improved the reliability and unifor-mity of catalyst production. It also permitted the polypropylene manufacturer to focus its expertise on improving polymer produc-tion. One of the first manufacturers in the Americas was Stauffer Chemical Company, dating from the late 1950s. Stauffer became involved when it serendipitously produced titanium trichloride during trials to manufacture titanium sponge, and it recognized the emerging market for propylene polymerization catalysts. In com-mercial operations excess titanium tetrachloride was heated with aluminum metal in large reactors, the resulting solids stripped of remaining $TiCl_4$, and the dry solid milled in temperature controlled paddle levigators (Figure 4.4) with steel balls [13]. The levigator design was more efficient for mechanical activation than conven-tional ball mills. After milling, the catalyst was classified by siev-ing and stray metal particles from the intensive levigation were removed by magnetic filters.

Incremental improvements in catalyst activity and stereospecific-ity were realized by extraction processes to remove $AlCl_3$ or by add-ing an organic compound containing a heteroatom, usually a Lewis

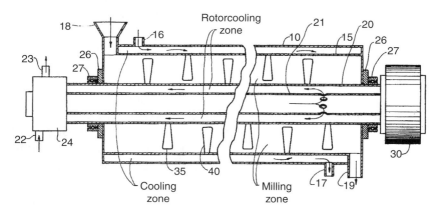

Figure 4.4 Stauffer Chemicals levigator for milling $TiCl_3$. Vertical paddles 35, 40 attached to a rotating central shaft 20 impact comixed $TiCl_3$ and steel balls in the milling zone. Source: A. Schallis, US Patent No. 3688992.

base, during the milling operations. The latter was initially more successful. These "third component" catalyst types also include variations where the organic compound was added with the aluminum alkyl cocatalyst. Many additives were explored, so that it seemed everything on the laboratory shelf was tried at some point. The patent literature describes a variety of aliphatic and aromatic ketones, esters, amines, ethers, thioethers, phosphorus compounds, amino-phosphorus compounds, siloxanes, etc. capable of incrementally improving stereoselectivity and/or activity [14, 15, 16, 17, 18, 19, 20, 21]. Typically, these were incorporated at less than 10% into the titanium trichloride, sometimes with final hydrocarbon washing to remove excess additive. Some additives had drawbacks; they affected polymer color, created flammability issues, carcinogenicity concerns, or were capable of creating peroxides. One commercially successful third component was camphor, which was capable of raising isotacticity to ~ 96 wt% [22]. All of these catalysts, based on titanium trichloride produced outside the polymerization reactor, with and without an added organic component, are called 1st generation catalysts.

First generation catalysts drove the global commercialization of polypropylene. The relatively modest activity and marginal stereospecificity required slurry processes in inert hydrocarbons to extract the atactic polymer, and alcohol or other post treatment to extract catalyst residues from the isotactic polymer. Slurry processes are further discussed in Chapter 8. The extracted sticky atactic polymer found use in adhesives, but its production outstripped early industrial requirements. So it was burned for fuel value, or landfilled.

4.4   Second Generation ZN Catalysts

Industrial chemists recognized that first generation catalysts had limits. Activity was too low to omit removing catalyst residues, and polymer isotacticity was still less than desired. This effectively limited process design to hydrocarbon slurries with post reaction catalyst neutralization processes. The search for improved performance continued, and it was rewarded when Solvay discovered that ether extractions were very effective to improve performance [23, 24]. Solvay's US patent No. 4210738, with a 1971 priority filing, describes that " by using complexing agents in a particular manner in the preparation of catalytic complexes based on $TiCl_3$, solids are obtained which have a large surface and which constitute catalytic systems having substantially higher activity and very good stereospecificity" [24]. The basic Solvay catalyst recipe is shown in eqs. (4.2) – (4.4):

$$TiCl_4 + Et_2AlCl \rightarrow TiCl_3 \cdot \tfrac{1}{3}(AlCl_3) + \tfrac{1}{3}Al_2Et_3Cl_3 + \tfrac{1}{2}C_2H_6 + \tfrac{1}{2}C_2H_4$$
$$Et = C_2H_5 \qquad (4.2)$$

$$TiCl_3 \cdot \tfrac{1}{3}(AlCl_3) + xsR\text{-}O\text{-}R \rightarrow TiCl_3 + (AlCl_3).R\text{-}O\text{-}R$$
$$R = C_nH_{2n+1} \qquad (4.3)$$

$$TiCl_3 + xs\ TiCl_4 \rightarrow \text{alkane wash} \rightarrow \text{activated } TiCl_3 \qquad (4.4)$$

The reduction with diethylaluminum chloride precipitates titanium trichloride, and spherical particles can be formed by controlling the reaction conditions. This constitutes an important advance. Due to the replication phenomenon, as described in Chapter 3, spherical catalyst particles produce spherical polymer particles. This enhances the flowability of the polymer, which has substantial practical significance. Packing efficiency increases, permitting greater polymer production per unit volume. In-process mixing is improved and higher particle density permits higher fluidization gas velocity without entrainment. Spherical polymer particles also increase polymer bulk density. Equipment remains cleaner during transfer operations, pneumatic conveyance is more reliable, and in storage silos the risk of bridge formation is reduced.

The chemical reduction with DEAC, eq. (4.2), generates approximately equal amounts of ethane and ethylene and produces a solid with significant amounts of aluminum chloride and aluminum alkyls coordinated to the surface. These are largely removed by extraction with an excess of an aliphatic ether, eq.(4.3), leaving behind some surface bound ether. The latter is then removed by complexation with $TiCl_4$, eq. (4.4), which also effects conversion of the crystallites into the delta form. Washing with a volatile hydrocarbon removes excess $TiCl_4$, yielding the final catalyst. The net result of this process is to produce a microporous form of $TiCl_3$ with a spherical morphology, high surface area and a narrow particle size distribution. The catalyst particle resembles a microporous sponge. These catalysts are markedly improved compared to the 1st generation, resulting in a five fold activity increase and about 2–3% increase in stereospecificity [25]. An example of a Solvay-type

Table 4.2 Relative polymerization performance of 1st and 2nd generation titanium trichloride [27].

Titanium Trichloride Type	Generation	Polymerization Activity g/g catalyst	Polymer Isotactic Index Wt %
Alpha	1	50	82
Beta	1	600	76
Gamma	1	100	86
Delta	1	1100	90
Delta + Camphor	1	1300	93
Solvay (Delta)	2	5000–10,000	96–98

Test conditions: ~ 142 psig C_3H_6 in heptane diluent, 3 psig H_2, 70°C, 4 hrs.

catalyst synthesis procedure may be found in Chapter 9. Later versions of the Solvay catalyst employed TEAL instead of DEAC in combination with an external donor to further increase activity and reduce residual chloride in the polymer [26]. These catalysts have the advantage of being less sensitive to catalyst poisons than supported 3rd or 4th generation catalysts, due to the higher titanium content in the former.

Solvay-type catalysts facilitated the development of the "bulk" polymerization process in which liquid propylene monomer replaced hexane or heptanes. The latter hydrocarbons were essential solvents for extracting atactic polymer, but the high stereospecificity of Solvay catalysts permitted using propylene as both monomer and as a weak solvent for extraction of a small amount of atactic polymer. A performance comparison of first and second generation catalysts is shown in Table 4.2 [27].

4.5   Third Generation ZN Catalysts

During the development of first and second generation catalysts work began on depositing $TiCl_3$ onto a support to increase the available Ti centers on the catalyst surface. Early examples included the use of silica, titania, alumina, and polypropylene itself. These were unexceptional with respect to yield of polymer per gram of catalyst [28, 29]. However, in the late 1960s Montedison and Mitsui researchers, working independently, found that activated $MgCl_2$ was a very effective support for forming a highly active catalyst in combination with titanium tetrachloride [30, 31, 32, 33]. Early patents included reference to zinc halides and manganese halides, but few examples were provided, and these salts proved ineffective. The unique properties of $MgCl_2$ have been ascribed to the similiarity in the radius of Mg^{+2} ion, 0.65Å, vs. Ti^{+4}, 0.68 Å, the similarity in bond lengths of Mg-Cl = 2.57 Å vs Ti-Cl = 2.51 Å, and isomorphism of $MgCl_2$ and $TiCl_3$ lattices [4].

Activation was accomplished by several routes, including milling $MgCl_2$, reactive dealcoholation of alcohol adducts of $MgCl_2$, and chlorination of alkylmagnesium or alkoxymagnesium compounds. Examples are given below:.

$$MgCl_2 \cdot nEtOH + xsMCl_4 \rightarrow MgCl_2 + nEtOMCl_3 + nHCl$$
$$M = Si, Sn, Ti; Et = C_2H_5 \tag{4.5}$$

$$MgR_2 + 2SiCl_4 \rightarrow MgCl_2 + 2RSiCl_3 \tag{4.6}$$

$$Mg(OEt)_2 + 2MCl_4 \rightarrow MgCl_2 + 2EtOMCl_3 \tag{4.7}$$

Reaction takes place in an inert hydrocarbon or in excess halogenating agent as solvent. In all cases, a high surface area support is formed, with small primary crystallites, on the order of 100 Å. The crystalline form, $\delta MgCl_2$, is disordered, creating many sites for binding titanium halides. When the reaction conditions are tightly controlled it is possible to get narrow particle size distribution, spherically shaped supports composed of agglomerated primary crystallites forming a microporous matrix. Hot filtration of the catalyst support from the halogenating agents is important in preventing reprecipitation of the byproducts.

The first application of supported Ziegler-Natta catalyst technology was for polyethylene because polypropylene isotacticity was initially poor [34]. Nevertheless, early patent claims still encompassed polypropylene: "a catalyst-forming component for use in preparing catalysts for the polymerization of olefins and which is the product obtained by contacting a titanium tetrahalide with an anhydrous magnesium dihalide in the active form" [30,31]. Applying a lesson from the use of third components in titanium trichloride catalysts, similar "electron donor" or "internal donor" compounds were added to the support, isotacticity improved dramatically, and the first practical third generation catalysts were realized [35]. Catalyst activity doubled versus Solvay catalysts, with highly acidic titanium chloride content replaced in large part by lower acidity magnesium chloride. Just as significantly, cocatalyst DEAC, typically used with $TiCl_3$ catalysts, was replaced with TEAL. This further reduced the highly acidic chloride in the polymer. Polymer isotacticity was somewhat lower, at about 95%, and decreased further when hydrogen was used for molecular weight control, which was a significant limitation. In certain cases a mixture of TEAL and DEAC gave the optimal balance of activity and isotacticity, at the expense of higher chloride residues [36].

Applying the principle that the donor compound functioned, in part, by poisoning atactic catalyst sites, it was discovered that adding an "external" donor into the polymerization reaction improved isotacticity, and that the best external donor depended upon the internal donor used in the catalyst synthesis. In general, monoesters were employed most successfully, and one particularly effective combination was to pair an ethylbenzoate internal donor with ethyl anisate or methyl-p-toluate external donor [37, 38, 39, 40]. Isotacticity could be maintained at ~ 90–95% depending upon the amount of hydrogen used to reduce molecular weight. The use of aromatic ester external donors imparted a faint sweet smell to the polymer.

4.6   Fourth Generation ZN Catalysts

With the development of third generation catalysts, the two primary goals of eliminating removal of catalyst residues and atactic polymer had been nearly achieved. Then, in 1981–1982 Montedison and Mitsui Petrochemicals developed supported catalysts that paired new internal and external donors: bifunctional phthalate esters and organosilylethers, respectively [41, 42, 43, 44]. TEAL remained the cocatalyst, with the added advantage that it was less reactive with organosilylethers than with aromatic esters. Catalyst activity increased again to about 50 kg/g, and polymer isotacticity reached 98%. Polymer odor was reduced due to the lack of aromatic monoesters. Activity was further enhanced by employing mixtures of $TiCl_4$ and an aromatic solvent, preferably toluene, as the extractant for support activation, and repeating the extraction several times [45]. With such a methodology catalyst activity could exceed 100 kg polymer/g catalyst [46]. Activity did not correlate directly with titanium content, the most active catalysts containing < 2 wt% Ti. Even so, kinetic investigations suggest the concentration of active sites is only a few percent of the titanium present [47, 48].

One of the consequences of the high activity 4th generation catalysts was that the heat of polymerization during the initial phase of polymerization could overheat the catalyst particle, fracturing it prematurely, and causing softening of the polypropylene particle. This resulted in several deleterious effects: a loss of activity, reduced stereospecificity, poor polymer morphology, and agglomerated particles. Therefore, methods were developed for moderating the polymerization rate for the first 10–100 g PP/g catalyst, a process called prepolymerization [49]. New developments in this area continue for both 4th and 5th generation catalysts [50]. Catalyst prepolymerization is discussed in section 3.6.

4th generation catalysts have an improved temperature response profile; in contrast to third generation catalysts, higher polymerization temperature increases activity without substantially degrading isotacticity. Similarly, increasing H_2 increases catalyst activity with only a minor effect on total isotactic index, smaller than for 3rd generation catalysts at equivalent MFRs.

During the course of the these developments, lessons learned in controlling the particle size distribution and porosity of Solvay-type catalysts, and first applied in third generation catalysts were more consistently used to control the morphology of $MgCl_2$ supports [51]. Milling the support as a form of activation was completely abandoned in favor of either physical or chemical methods to produce spherical particles, followed by chemical extraction steps with $TiCl_4$ to produce a highly porous $MgCl_2$ support. One particularly favored chemical route was controlled precipitation of alcohol adducts of $MgCl_2$ followed by extraction of the alcohol by reaction with metal halides, especially $TiCl_4$ or $SiCl_4$. Physical methods included preparation of a molten solution of $MgCl_2$ in an alcohol, emulsification in a mineral oil, and quenching in a cold hydrocarbon. The emulsification method determines the particle size distribution of the support droplets, and the quenching freezes the droplet shape into solid particles. A variety of physical methods and equipment have been developed to form the desired dispersion, including high speed agitation, turbulent flow through a pipe, and spray chamber methods [52]. One device is shown in Figure 4.5.

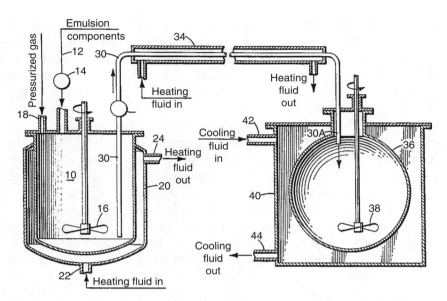

Figure 4.5 Apparatus to form spherical support particles [52] (b). An emulsion of a magnesium compound is formed in 10, the emulsion flows through pipe 34 under turbulent conditions, and the emulsion in quenched in 36. Source: M. Ferraris, F. Rosati, US Patent No. 4469648.

Through different mechanical or chemical precipitation conditions it was possible to produce spherical or polyhedral catalysts from 10–100 microns with narrow particle size distributions, ultimately yielding resin with 0.3–4.0 mm diameter [53, 54]. A representative comparison of catalyst particles and polymer particles is shown in Figure 4.6. The larger polymer particles enabled the development of multireactor processes for impact copolymers. These processes incorporated rubbery impact copolymer phases "buried" within the homopolymer sphere, without leading to particle agglomeration from stickiness. One illustrative example is low blush polypropylene, which gives low whitening when stressed due to the incorporation of rubbery microdomains. It is produced sequentially by mulitple reactors that incorporate into a single polymer particle, isotactic polypropylene, a propylene-ethylene copolymer, and an ethylene-α-olefin copolymer phases [55].

Research efforts to control morphology also included synthesizing catalysts consisting of a thin layer of $MgCl_2$ coated onto the pores of a preformed spherical silica support. This may be done by reaction of an alkylmagnesium precursor with spherical silica and subsequent

a) Supported Catalyst, [57] ~25 μ diameter   b) Polypropylene, ~700 μ diameter

Figure 4.6 Comparison of supported catalyst particles and resultant polypropylene particles. Reproduced by permission of Elsevier Ltd [56].

halogenations, or alternatively dealcoholation of a $MgCl_2$ solution in the presence of silica [58, 59, 60]. Generally, these catalysts have proven less active due to the inert silica diluent in the catalyst [61].

The fourth generation catalysts finally attained the performance needed to fully realize processes that did not incorporate a diluent, nor solvents for catalyst deashing. Therefore, industrial developments moved away from hydrocarbon slurry processes to bulk propylene, gas phase, and tandem or multiple reactor processes (these are discussed in Chapter 8). This both simplified the homopolymer production (no solvents), and enabled the production of new impact copolymers which would be too soluble for production in a hydrocarbon slurry or liquid propylene process. This was made possible by fourth generation catalysts having three key properties: high isotacticity, high activity, and controlled particle morphology. These catalysts are presently used for the majority of global polypropylene production for injection molding, films, and fibers, including most copolymer grades. The improvements in stiffness (high isotacticity) and toughness (high impact resistance) have expanded the available performance range and driven up global demand for polypropylene resins.

4.7   Fifth Generation ZN Catalysts

The achievements of the fourth generation of catalysts fulfilled the essential requirements for simplified bulk and gas phase processes:

high activity and high isotacticity. This encouraged researchers to refocus their targets toward catalysts tailored for special grades to extend the range of available polymer properties.

In assessing the development of 3rd and 4th generation catalysts, it is clear the use of titanium, MgCl$_2$, and triethylaluminum are "fixed", i.e., these components are essential to the catalyst. There is little current ZN research directed toward replacement of those components [62]. Instead, researchers recognized that the organic donors were the vehicles for further catalyst modification. The progress described below is the outcome of research from varying the structures of the internal and external donors to customize catalyst performance toward specific ends. Three areas of development are discussed below: narrow MWD, broad MWD, and modified polypropylene microstructure.

In achieving almost perfect isotacticity from 4th generation catalysts a certain breadth of molecular weight distributions was realized, typically PDI = 4–7. Now researchers looked to extend this range in both directions. Continuing their leadership in supported catalyst technology, Montedison researchers in 1988 identified a class of substituted 1,3 diethers that permitted elimination of the external donor from the catalyst system [63, 64, 65, 66, 67]. Examples include 2,2,diisobutyl-1,3-dimethoxy propane (I) and 9,9 bis(methoxymethyl)flourene (II):

(I)   (II)

Very high activities and total xylene insolubles of 96–98% were reached without an external donor, and with an external donor 99% TXI was possible [68]. The superior activity has been attributed to an increase in the concentration of active centers, as opposed to a change in rate constants [48]. Very low stereo defects permit the melting temperature of the polypropylene to reach 165°C. However, the main advantage of these catalysts is that they are highly responsive to H$_2$, having over 20 times the sensitivity compared to 4th generation catalysts using phthalates. This enabled making very high melt flow rates up to 1000 dg/min^{-1} and narrow molecular weight distributions, PDI ~3–5. This has been attributed to more frequent 2,1 insertions, which subsequently result in chain scission with hydrogen [69]. Narrow MWD should also be promoted by a more narrow range of catalyst site structures. Accordingly, one study demonstrated that diether internal donor promoted formation of $MgCl_2$ crystallite formation with only 120° edge angles, whereas ethyl benzoate or alkyl phthalates lead to $MgCl_2$ crystallites with both 120° and 90° edge angles [70]. Hence, the diether would promote more uniform catalyst environments.

Applications include thin-wall injection molding and spun fibers. In many cases 5th generation catalyst technology eliminates the need for peroxide initiated visbreaking to achieve narrow MWDs and high MFRs. Oligomer content was also reduced, improving organoleptic properties. The high catalyst activity reduces the amount of catalyst neutralizers which generate smoke during processing.

A second family of fifth generation catalysts, developed by Basell, is based on succinates as internal donors. Whereas diethers were developed to achieve narrow MWDs, the succinate catalysts broaden molecular weight distribution, PDI~5–15 [71]. This is realized in some instances by incorporating succinate and phthalate esters into the same catalyst, and may be rationalized as creating a broader range of catalytic environments about titanium [72, 73, 74, 75]. Examples of succinates include rac-diethyl 2,3, diisopropylsuccinate (III) and dibutylsuccinate (IV), shown below.

(III)   (IV)

A third family of 5th generation catalysts also derives from modifying the donors. Certain pairs of external and internal donors, specifically n-octyl(methyl)dimethoxysilane and 2-octyl-2-methyl-1,3-dimethoxypropane, have been identified as able to modify the microstructure of the polypropylene to produce high amounts of stereoblock segments. Stereoblocks are sections of polymer which are intermediate in stereospecificity, neither highly isotactic, nor atactic. The stereoblocks confer softness to the resin without stickiness. When used in combination with multi-reactor technology, new very soft grades of polypropylene are produced. These grades maintain their thermal and optical properties for applications requiring transparency, softness, high tenacity and low temperature resistance such as packaging films or flexible closures [76, 77].

4.8   ZN Catalysts for Atactic Polypropylene

The early Ziegler-Natta catalysts produced relatively large amounts of atactic polypropylene which was separated from the isotactic polymer by dissolution in hydrocarbons used as polymerization solvent. The atactic polymer included a certain fraction of low molecular weight isotactic polymer which could crystallize, low molecular weight oils and high catalyst residues. Production of this tacky material exceeded its market use in adhesives, sealants, caulking compounds, and roofing formulations. Therefore, a portion of it was burned for fuel value or disposed of in landfills [78]. With succeeding generations of catalysts the production of atactic polypropylene decreased as a fraction of the total polymer and finally reached the point where it was no longer separated from the isotactic polymer. However, the market for adhesives continued to develop and demand for atactic feedstock increased. So, at a certain point a shortage of atactic polymer developed, and it became economically interesting to develop catalysts to produce atactic polypropylene. Initially, this included recovering atactic polypropylene that had been previously landfilled. The polymer was strip mined, washed and sold [79].

The first catalysts combined $TiCl_4$, $MgCl_2$ and an alkylaluminum compound in the absence of any internal or external donor [80]. No effort was expended at controlling catalyst morphology because atactic polypropylene must be produced as a dissolved or liquid product. Catalyst activity was ~ 1000 lb/lb in a hydrocarbon polymerization. The liquid product, after stripping the hydrocarbon, contained minimal crystalline polymer and oil, and ~ 0.1 wt% catalyst. By adding a minor portion of ethylene into the polymerization step the adhesive properties could be optimized. Accordingly, the nomenclature amorphous polyalphaolefins (APAO) began to be applied as a more descriptive term. New solution processes were developed, such as by El Paso Products, and a third comonomer, butene, further tailored the APAO properties [81]. The catalyst consisted of a comilled mixture of $MgCl_2$, $AlCl_3$, and $TiCl_4$. Activity in excess of 50,000 lb/lb was reached in combination with a cocatalyst mixture of TEAL and DEAC. It was recognized that the temperature maximum imposed by the stability limits of isotactic catalysts, about 90–100°C, did not restrict APAO production. Accordingly, the catalyst and cocatalyst were applied in a continuous tubular process at 150°C–230°C, in the absence of hydrocarbon solvent, and with short residence times of a few minutes. A molten APAO product is discharged. In order to further modify the APAO tensile properties, knowledge from isotactic processes was applied, and external donors were added into the cocatalyst mixture [79,82].

With the further development of APAO, new resins were produced with controlled crystallinity intermediate between APAOs and typical isotactic resins. These resins could produce flexible forms with low stickiness, and were termed FPOs, or flexible polyolefins. This was achieved by milling $MgCl_2$, $AlCl_3$, and $TiCl_4$ with two internal donors, an organosilylether such as cyclohexylmethyldimethoxysilane (CMDS), and a hindered amine, such as 2,6 lutidine [83]. The former was found increase catalyst activity, while the latter minimized low molecular weight polymer. These catalysts were employed with TEAL and CMDS external donor to produce the FPO. Still other recent developments have focused on utilizing $TiCl_3.1/3AlCl_3$, indicating that the early ZN catalysts are still finding some commercial use for certain grades of polypropylene [84].

From the discussions above, the reader will recognize that a family of Ziegler-Natta catalysts has evolved, each member with different purposes and attributes. Research was first directed toward achieving highly active catalysts producing highly isotactic polymer. Attention then turned to catalysts for making highly atactic polymer or APAOs. Finally, the gap between the polymers was bridged by catalysts designed to make polymers with intermediate crystallinity, the FPOs. Together, a kind of continuum in polymer properties is achieved. The catalyst recipes share much in common with this continuum, in particular the judicious incorporation of specific classes of organic compounds to achieve the desired ends.

4.9 Metallocenes and Other Single Site Catalysts

Concurrent with the development of Ziegler-Natta catalysts for polypropylene, over the last several decades there has been intensive research into the synthesis of discrete, molecular organometallic polymerization catalysts. Historically, these catalysts originated as a family of metallocenes, but many other structurally complex organometallic catalyst structures have subsequently been synthesized. Collectively, they are called single site catalysts (SSCs). In the case of metallocenes, these catalysts have been applied to commercially produce a wide range of polymers, including polypropylene.

Philosophically, the development of SSCs is a counterpoint to the ill-defined nature of heterogeneous Ziegler-Natta catalysts, and SSCs compete with those catalysts for commercial applications. However, the continuous progress in ZN catalysts has influenced the development of SSCS to focus upon complimenting ZN catalysts by emphasizing the production of new microstructures that extend the range of accessible polymer types and physical properties. Chemically, many ZN and SSC catalyst systems are not entirely compatible with each other. Nevertheless, a commercial polymerization plant may employ both types in campaign mode in a common reactor, and transition between resins is a concern. Therefore methods have been developed to changeover production between catalyst types with a minimum of interferences [85]. The evolution of SSCs does not fall neatly into generations of catalysts; it is a multifaceted area and discussed separately in Chapter 6.

4.10 Cocatalysts for ZN Catalysts

During the initial stages of ZN catalyst development, many metal alkyl compounds were investigated as alkylating agents to initiate polymerization at the metal center. These included the pyrophoric alkyls of the alkali metals, alkaline earths and Group III metals. Sodium and potassium alkyls are limited by their instability and poor solubility in hydrocarbons. Berylium is too toxic for practical use. Gallium and indium are relatively expensive. So, investigations focused on lithium, magnesium, boron, aluminum, and zinc. Performance and availability favored work on aluminum. A particular advantage of aluminum is its tunability in a hydrocarbon medium. As a trivalent metal there are many ligand combinations possible, and the alkylaluminum halides, alkoxides and amides are all miscible in hydrocarbons. By contrast, boron alkyls proved relatively inactive. Lithium has only one valency, limiting the versatility of ligands, and lithium alkyls are expensive. Magnesium has two valences; but its simple dialkyls

(C$_1$ to n-C$_4$) are insoluble in hydrocarbons. However, several mixed magnesium alkyls such as n-butyl(ethyl)magnesium (BEM) are highly soluble in hydrocarbon. BEM is widely used today to produce catalysts for polyethylene [86]. In contrast, alkylmagnesium halides (Grignard reagents) require aprotic Lewis bases (typically ethers) as cosolvents, and Lewis bases are generally detrimental to polymerization activity. Zinc, also with only two valencies, tends to form preferentially dialkylzinc compounds from alkyl halides, with concomitant production of an equivalent of zinc dihalide [87]. Alkylmagnesium compounds would find eventual use as precursors to MgCl$_2$ supports, and diethylzinc has found some limited use in molecular weight control [88]. However, it is the alkylaluminum family that is universally used for ZN cocatalysts for polypropylene. Triethylboron, used in small amounts vs TEAL, has been found useful to increase polymerization yield and broaden MWD [89].

The utility of aluminum alkyl compounds derive from their reducing and alkylating power, their ability to scavenge polar impurities in propylene and their complexation of external donors to moderate donor properties. Smaller ligands maximize the moles per pound of aluminum alkyl, so triethylaluminum (TEAL) and diethylaluminum chloride (DEAC) are almost always the reagents of choice. The TEAL and DEAC manufacturing routes are economical (see Chapter 5), much more so than trimethylaluminum. Triisobutylaluminum is little used; it usually provides no particular process advantages, is only available in lower purity than TEAL and DEAC, and contains a lower aluminum content than TEAL (see section 5.2.2).

The pyrophoric nature of aluminum alkyls, *i.e.* their reactivity with oxygen and moisture, is consistent with their utility for alkylation of transition metals and reactivity with catalyst poisons. It derives from the lability of trivalent aluminum, which, as a Lewis acid, can form tetravalent coordination, and from the relative bond strengths of the Al-C and Al-O bonds:

bond dissociation energy [90]
(kJ/mol)
Al-C   255
Al-O   512

Some basic properties of DEAC and TEAL are shown in Table 4.3 [91].

Table 4.3 Properties of DEAC and TEAL.

Property	Diethylaluminum Chloride	Triethylaluminum
Acronym	DEAC	TEAL
CASRN	96–10–6	97–93–8
Form	Clear, colorless liquid	Clear, colorless liquid
Molecular Formula	$AlClC_4H_{10}$	AlC_6H_{15}
Molecular Weight	120.56	114.17
Structure	Dimer	Dimer
Aluminum content, wt% (theoretical)	22.4	23.6
Chorine content, wt% (theoretical)	29.4	0
Melting Point, °C	-85	-52
Boiling Point, °C @ 760 mm Hg	214	186
Vapor Pressure, mm Hg/°C [100, 101]		
0	0.0177	0.00134
10	0.0461	0.00466
20	0.111	0.0147
50	1.04	0.283
100	17.1	11.7
Density g/mL @°C	0.961@25°C	0.832@25°C
Viscosity, cp @°C	1.4@30°C	2.6@25°C
Specific Heat, cal/g @57°C	0.410	0.532
Heat of Hydrolysis, cal/g	885	1104
Heat of Combustion, kcal/mole	856	1220
Nonpyrophoric Limit, wt% in hexane [102]	13	12
Solubility in hydrocarbons	Miscible	Miscible

DEAC dominated the early development of ZN catalysts with TiCl$_3$, giving a better combination of activity and isotacticity than TEAL. With the development of supported TiCl$_4$ catalysts the higher reducing power of TEAL universally gave better performance. Further, TEAL does not contain chloride. The latter contributes to the corrosivity of the polymer due to release of HCl upon contact with moisture. So, TEAL use, in place of DEAC, also reduces the required amount of neutralization additives added to the polypropylene in post reactor.

TEAL interacts with the external donors employed in 3rd and 4th generation catalysts. In the case of aromatic esters, a Lewis acid/base complex forms and then alkylation occurs at the carbonyl center:

methyl p-toluate (MPT) (4.8)

Further alkylation can occur in the presence of additional TEAL [92, 93]. The resulting products are not as effective as external donors, and this may contribute to the relatively high amounts of MPT and similar aromatic esters (TEAL: ester = 4:1) used in 3rd generation catalyst systems.

In 4th generation catalysts alkoxysilanes are used. Dialkoxydialkylsilanes form relatively stable 1:1 complexes with triethylaluminum under typical polymerization conditions, based on NMR investigations [94]. Alkyltrialkoxysilanes have a greater tendency toward ligand exchange, forming dialkyldialkoxysilanes, which still function as effective external donors [95]. This is shown in equations (4.9) and (4.10).

CMDS (4.9)

$$(4.10)$$

It is the interaction of these complexes with the catalyst surface that generates isospecific sites on the catalyst surface. This may explain why organosilanes are used in much lower amounts than aromatic esters as external donors, Al:Silane = 20:1. Thus, triethylaluminum acts as a moderating agent for the external donor. There is some evidence that TEAL may be structurally incorporated into the active site, such as through bridging structures [96, 97]. In contrast, complexed esters rapidly exchange with the external solution [98].

Like many industrial chemicals, TEAL and DEAC contain small amounts of impurities (discussed in Chapter 5). These were not a particular issue during the early stages of catalyst development because the catalyst activities were low and the Al/Ti ratio was ~4/1. Purity became increasingly important for TEAL as the catalyst activity increased. Al/Ti increased to ~50/1 and DEAC usage declined. Also, with the conversion from 3rd generation to 4th generation catalysts external donor usage was reduced fivefold on a molar basis, and aluminum alkyl impurities began to noticeably affect polymerization results. It was found that low levels of diethylaluminum hydride, Et_2AlH, reduced catalyst stereospecificity. The Et_2AlH arises from incomplete reaction in the synthesis of TEAL:

$$(4.11)$$

The Et_2AlH was found to rapidly react with organosilanes in a ligand exchange reaction. The resulting silane proved to be an ineffective donor [99].

$$R_2Si(OCH_3)_2 + Et_2Al\text{-}H \rightarrow R_2SiH(OCH_3) + Et_2Al\text{-}OCH_3 \quad (4.12)$$

As a consequence, aluminum alkyl manufacturers improved TEAL quality to reduce hydride to vanishingly low levels.

TEAL and DEAC are used as neat liquids in polypropylene plants, despite the lower pyrophoricity of hydrocarbon solutions. Accordingly, special handling procedures are used, which are described in Chapter 6.

Approximately 40–200 moles TEAL/mole Ti is employed with 4th or 5th generation catalysts, corresponding to 2–10 lb TEAL/lb catalyst for a nominal 2% Ti catalyst. This permits the TEAL to also function as a scavenger for propylene impurities:

$$Et_3Al + \tfrac{1}{2}O_2 \rightarrow Et_2AlOEt \tag{4.13}$$

$$2Et_3Al + H_2O \rightarrow Et_2Al\text{-}O\text{-}AlEt_2 + 2EtH \tag{4.14}$$

$$Et_3Al + CH_3OH \rightarrow Et_2AlOCH_3 + EtH \tag{4.15}$$

$$Et_3Al + CO_2 \rightarrow Et_2Al\text{-}O\text{-}C(=O)\text{-}Et \tag{4.16}$$

$$Et_3Al + HC\equiv CH \rightarrow Et\text{-}CH=CH\text{-}AlEt_2 \tag{4.17}$$

$$Et_3Al + (CH_3)_2NH_2 \rightarrow Et_2Al\text{-}N(CH_3)_2 + EtH \tag{4.18}$$

$$Et_3Al + COS \rightarrow Et_2Al\text{-}O\text{-}C(=S)Et \text{ [103]} \tag{4.19}$$

Although TEAL is a Lewis acid it is an ineffective scavenger for carbon monoxide. Presumably this is due to TEAL's dimeric structure (see Figure 5.1 for analogous TMAL dimer) which reduces Lewis acidity [104].

4.11 Kinetics and ZN Catalyst Productivity

By its nature, the kinetics of heterogeneous catalysis is complex. For propylene polymerization this is further complicated by several types of active sites and the presence of additional components in the system: cocatalyst, external donor, hydrogen, and catalyst poisons. Nevertheless, several qualitative statements can be made regarding kinetics and the overall catalyst productivity in modern supported catalyst systems, and the reader is referred to other sources and references therein for more detailed studies [1,4,105, 106, 107]. The following factors are discussed briefly: time, catalyst

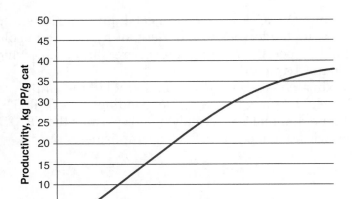

Figure 4.7 Productivity vs. Time for a typical supported catalyst.

concentration, Ti concentration, propylene purity and concentration, cocatalyst, external donor, and hydrogen.

Time
A prototypical shape of a polymerization curve versus time is shown in Figure 4.7, and is highly dependent on the specific supported catalyst. Following prepolymerization, the productivity reaches an approximately constant rate and then decreases slowly. The gradual decrease in rate is usually ascribed to a slow deactivation of active sites by poisons, over-reduction of titanium to the +2 valence state, and slow side reactions with the donors present. Despite the fact that the catalyst particles become embedded in the growing polypropylene matrix, it is not believed that diffusion limitations are responsible for the gradual reduction in productivity with time [108, 109].

Temperature
Supported catalysts usually have a maximum polymerization rate in the 70–85°C region. Above that rates and stereospecificity often decrease, although specific catalyst types have been prepared to operate in supercritical propylene above 90°C (see section 8.7).

Catalyst
The concentration, or amount, of catalyst directly impacts the polymer production rate per unit volume of reactor. This is not to be

confused with the polymer production rate per weight of catalyst, which is not affected. In general, kinetics information is used to get the highest productivity from the catalyst system. However, from a practical standpoint, increasing the polymerization rate requires increased heat removal capacity from the reactor. Once that heat removal limit is reached, it is a barrier to further improvements in kinetics.

Titanium

The concentration of titanium on the catalyst does not have a direct relationship to the rate of polymer production. That is, there is no simple generalization that productivity benefits from higher titanium in the catalyst. This is consistent with the understanding that only a small fraction of the titanium is catalytically active [48]. Some of the highest activity supported catalysts reported in the literature have relatively low titanium content. High productivities are typically reported for catalysts with about 1–4 wt% titanium.

Propylene Purity and Concentration

Propylene purity is an important consideration in catalyst productivity, and its purification is an important aspect of commercial processes. Many impurities are scavenged by aluminum alkyls, so, in some cases increasing the aluminum alkyl concentration can improve kinetics. This, of course, depends on the poison's rate of reaction with the aluminum alkyl versus its rate of reaction with the catalyst. In particular, with semi-batch slurry or gas phase polymerizations the poisons are introduced continually with the propylene, and one may expect that this will have an accumulative effect on polymerization rate.

The concentration of propylene directly affects kinetics, with a first order effect. Thus, increasing propylene concentration in slurry or gas phase will give a proportionate increase in rate. Of course, bulk processes using neat liquid propylene are at the concentration limit.

Cocatalyst

Aluminum alkyl cocatalyst is generally used in large molar excess relative to the catalyst, and so, its role in polymerization kinetics is minor, depending on the extent of propylene impurities.

External Donor

The external donor is employed in a high enough concentration to assure that atactic polymer production is suppressed. Increasing concentration significantly beyond that point will generally begin to suppress catalyst productivity.

Hydrogen
Hydrogen has a distinct activating effect for supported catalysts, as much as an order of magnitude, and it eventually reaches a plateau point beyond which polymerization rate is unaffected, but polymer melt flow rate continues to increase [110]. The hydrogen is believed to function, in part, by reactivating sites that become inactive due to insertion errors.

4.12   Concluding Remarks

The development of ZN catalysis for polyolefins, especially polypropylene, has been the subject of intensive investigation for over half a century. Development has been primarily driven by industrial laboratories, with the investment in research programs justified by the profits derived from steadily growing demand for the resin. The industrial investigations have been extremely successful, dramatically impacting polymer plant design, production economics, resin quality and the variety of resins available. Work continues to this writing, with industrial scientists still at the lead. Understanding the history of ZN catalysis is essential to understanding the evolution of the polyolefin industry, and it is an important subject for all students interested in heterogeneous catalysis or polymers. A generational historical catalyst summary is shown in Table 4.4.

4.13   Questions

1. A propylene polymerization system is designed to consume 40% of the available propylene using a catalyst that gives 50,000 lb PP/lb catalyst. The catalyst contains 2.5 wt% titanium and 100TEAL/Ti. The propylene contains 1 ppm H_2O impurity. Assuming that TEAL reacts with H_2O as in eq. (4.14), calculate the fraction of TEAL that is theoretically consumed by the water. What conclusion do you draw?

2. An investigator claims to have developed an active polymerization catalyst based on iron chloride, TEAL, and an

Table 4.4 generations of Ziegler-natta propylene polymerization catalyst.

Generation	Support Type	Morph. Control	Internal Donor	External Donor	Cocat.	Activity kg PP/g cat	Isotac-ticity %	Polymer Morphology	PDI
0	βTiCl$_3$ (no support)	No	No	No	TEAL	0.1–0.2	65–80	Powder and flake	
1	δTiCl$_3$ (no support)	No	Lewis Bases	No	DEAC	1–2	90–95	Powder and flake	6–10
2	δTiCl$_3$ (no support)	Yes	Ethers	No	DEAC or TEAL	4–6	94–97	Spherical	7–10
3	MgCl$_2$	No	Aromatic esters	Aromatic Esters	TEAL	5–15	90–95	Typically powder and flake	5–8
4	MgCl$_2$	Yes	Aromatic Diesters	Organosilyl ethers	TEAL	30–60	94–98	Spherical, narrow psd	4–7
5	MgCl$_2$	Yes	Diethers	No	TEAL	100	96–99	Spherical, narrow psd	3–5, high MFRs
5	MgCl$_2$	Yes	Succinates	Organosilanes	TEAL	40–70	95–99	Spherical, narrow psd	5–15

aniline donor. Provide some advice to the investigator on the practicality of this catalyst system.

3. A hypothetical polypropylene plant still manufactures polymer by a batch process. The reactor uses catalyst A producing 2000 g PP/g catalyst.hr., and the polypropylene has a bulk density = 0.40 g/mL. The reactor operates until 35 bulk volume % is occupied. This takes 4 hours. Emptying the reactor and preparing for the next batch takes 1 hour. A Solvay catalyst S is offered. It produces 3000 g PP/g catalyst.hr with bulk density 0.50 g/mL. Quantify the advantages this catalyst offers in throughput to the plant.

4. Given that dilute hydrocarbon solutions of TEAL are below the nonpyrophoric limit, what reasons might influence the decision of polymer producers to use neat TEAL?

5. A chemist wants to investigate triisobutylaluminum, TIBAL, in place of TEAL as a cocatalyst for a fourth generation catalyst. Typical compositions are compared below. What advice would you give the chemist?

Table 4.5 TIBAL – TEAL comparison.

TIBAL	TEAL
96.6% triisobutylaluminum	95.0% triethylaluminum
0.2% tri-n-butylaluminum	4.9% tri-n-butylaluminum
0.1% other	0.1% other
2.6% isobutylene	0.0% isobutylene
0.5% hydride as AlH_3	0.02% hydride as AlH_3
mp= 0°C	mp= –52°C

6. Stickiness can cause polypropylene particles to agglomerate, and it is a surface phenomenon. Explain why a sphere is the preferred shape to minimize stickiness. Demonstrate why larger spheres are preferred over smaller spheres to minimize stickiness from the incorporation of impact copolymer phases.

7. Assume the heat capacity of polypropylene is constant at 0.4 cal/g/°C. Assume the heat of polymerization of propylene is constant, at 20 kcal/mole, and that the heat of fusion of polypropylene is 49 cal/g. Calculate the theoretical

adiabatic temperature rise of propylene polymerization. What is the practical implication of this calculation for high activity catalysts?

References

1. *Polypropylene Handbook*, 2nd edition, Nello Pasquini editor, Hanser Publishers 2005.
2. (a) *Polypropylene Handbook*, 2nd edition, Nello Pasquini editor, Hanser Publishers 2005., pp. 8–11; (b) Heinz Martin, *Polymers, Patents, Profits*, WILEY-VCH Verlag GmbH & Co., 2007.
3. (a) Avant Catalysts for Polyolefin Production, Doug Gibney, Lyondell-Bassell Industries, Feb. 25, 2008, (b) http://chemicals.ihs.com (c) A. Ho, 2010 APIC-CMAI Seminar, Mumbai, India, May 13, 2010, cmaiglobal.com.
4. Ziegler-Natta Catalysts, *Kirk Othmer Encyclopedia of Chemical Technology*, John Wiley and Sons Inc., vol. 26, pp. 502–554.
5. G. Natta, P. Pino, G. Mazzanti, Gazzanti, Gazz. *Chim. Ital.*, vol. 87, p. 528.,1957; G. Natta, P. Pino, G. Mazzanti, Gazzanti, P. Longi, Gazz. Chim. Ital.,vol. 87, p. 549–570, 1957.
6. In early 2011, tantalum and gadolinium metals prices exceed $200/lb, versus $3/lb titanium and $1/lb aluminum.
7. Toxic catalyst recipes did not go unexplored. See, for example catalysts employing mercury and lead found in Y. Tsunoda, S. Fujimoto, I. Aishima, Y. Kobayashi, US Patent No. 3065216.
8. G. Natta, P. Pino, G. Massanti, US Patent No. 3014018, filed Feb. 25, 1958.
9. H.J. Hagemeyer, M.B. Edwards, US Patent No. 3067183, filed Oct. 15, 1956.
10. E. J. Vandenberg, US Patent No. 3051690.
11. J. Boor Jr., Ziegler *Natta Catalysts and Polymerizations*, Academic Press, 1979, p 95.
12. E. Tornqvist, A.W. Langer Jr., US Patent No. 3032510, filed June 27, 1958.
13. A. Schallis, US Patent No. 3688992, filed April 15, 1970.
14. S. Wada, H. Oi, N. Matsuzawa, H. Nishimura, J. Sasaki, US Patent No. 3701763, filed Oct. 20, 1970.
15. H.D. Lyons, C.W. Moberley, US Patent No. 3210332.
16. A.D. Caunt, M.S. Fortuin, US Patent No. 4051307.
17. H. Schick, P. Hennenberger, G. Staiger, H.M. Tamm, US Patent No. 3951858.
18. Reference 9, pp. 112–115.
19. N. Matsuzawa, H. Oi, H. Nishimura, S. Wada, J. Sasaki, US Patent No. 4048415.
20. J. Hotta, M. Fujii, US Patent No. 4020264.

21. T. Shiomura, A. Ito, Y. Morimoto, T. Iwao, US Patent No. 4028481.
22. G.G. Arzoumanidis, US Patent No. 4124530.
23. J.P. Hermans, P. Henrioulle, US Patent No. 4210735.
24. J.P. Hermans, P. Henrioulle, US Patent No. 4210738 priority March 23, 1971.
25. A. Bernard, P. Fiasse, pp. 405–424, *Catalytic Olefin Polymerization*: Proceedings of the International Symposium on Recent Developments in Olefin Polymerization Catalysts, Tokyo, October 23–25, 1989, T. Keii and K. Soga editors, Elsevier, 1990.
26. J.P. Costa, S. Pamart, H. Collette, S. Bettonville, US Patent No. 6083866.
27. K. B. Triplett, The Evolution of Ziegler Natta Catalysts for Propylene Polymerization, Chapter 7, *Applied Industrial Catalysts*, Volume 1, B.E. Leach, editor, Academic Press, 1983.
28. J.P. Hermans, S. Bever, P. Henrioulle, US Patent No. 3769233. Priority Mar. 26, 1970.
29. T. Tomoshige, S. Honma, US Patent No. 3718635.
30. A. Mayr, P. Galli, E. Susa, G.DiDrusco, E. Giachetti, US Patent No. 4298718.
31. A. Mayr, E. Susa, E. Giachetti, US Patent No. 4476289.
32. A. Toyota, K. Odawara, N. Kashiwa, US Patent No. 4085276.
33. A. Toyota, N. Kashiwa, S. Minami, US Patent No. 4157435.
34. N. Kashiwa, US Patent No. 4071672.
35. A. Toyota, K. Odawara, N. Kashiwa, US Patent No. 4076924.
36. This cocatalyst system was applied, for example in: C.M.Selman, L.M. Fodor, US Patent No. 4379898; M. Takitani, S.Tomiyasu, K. Baba, US Patent No. 4525558.
37. U. Gianinni, A. Cassata, P. Longi, R. Mazzocchi, US Patent No. 4187196, Priority June 1971.
38. L. Luciani, N. Kashiwa, P.C. Barbe, A. Toyota, US Patent No. 4331561.
39. U. Gianinni, E. Albizzati, S. Parodi, US Patent No. 4149990. Priority Aug.1976.
40. L. Luciani, N. Kashiwa, P.C. Barbe, A. Toyata, US Patent No. 4226741.
41. M. Terano, H. Soga, M. Inoue, K. Miyoshi, US Patent No. 4547476.
42. N. Kashiwa, Y. Ushida, US Patent No. 4442276. Priority Feb 1982.
43. E. Albizzati, S. Parodi, P.C. Barbe, US Patent No. 4522930. Priority Feb, 1982.
44. V. Banzi, P.C. Barbe, L. Noristi, US Patent No. 4529716. Priority Sept.1982.
45. M. Terano, A. Murai, M. Inoue, K. Miyoshi, US Patent No. 4816433; T. Kataoka, H. Umebayashi, K. Goto, US Patent No. 5945366; M. Terano, H. Soga, M. Inoue, US Patent No. 5130284.
46. R.A. Epstein, W.T. Wallack, US Patent No. 7504353.
47. P.J.T. Tait, I.A. Jaber, A.J. Loontjens, pp. 11–28, *Catalytic Olefin Polymerization*: Proceedings of the International Symposium on Recent Developments in Olefin Polymerization Catalysts, Tokyo, October 23–25, 1989, T. Keii and K. Soga editors, Elsevier, 1990.

48. A.K. Yaluma, J.C. Chadwick, P.J. T. Tait, *Macromolecular Symposia*, vol. 260, pp. 15–20, 2007; A.K. Yaluma, J.C. Chadwick, P.J. T. Tait, *J. Polymer Sci.*: Part A: Polymer Chemistry, vol. 44, 1635–1647, 2006.
49. *Polypropylene Handbook*, 2nd edition, Nello Pasquini editor, Hanser Publishers 2005., p. 78.
50. See for example: (a) L.P. Leskinen, O. Tuominen, US Patent Application No. 2010/0184929A1; (b) T. Goto, K.Wakamatsu, S. Kumamoto, Shinichi US Patent No. 6645901; (c) G. Collina, O. Fusco, US Patent Application No. 2010/0292419A1; (d) G. Mei, M. Covezzi, S. Bertolini, US Patent No. 7652108; (e) G. Govoni, M. Sacchetti, G. Vitale, US Patent No. 6468938; (f) D.B. Morse, US Patent Application No. 2001/0051697A1.
51. M. Grun, K. Heyne, C. Mollenkopf, S. Lee, R. Uhrhammer, US Patent No. 7767772.
52. (a) G. Foschini, N. Fiscelli, P. Galli, US Patent No. 4111835; (b) M. Ferraris, F. Rosati, US Patent No. 4469648; (c) Y. Yang, H. Du, Z. Li, Z. Wang, Z. Tan, K. Zhang, X. Xia, T. Li, X. Wang, T. Zhang, W. Chen, X. Zheng, US Patent Application No. 2006/0003888.
53. R. Spitz, T. Soto, C. Brun, L. Duranel, US Patent No. 62324221.
54. P. Rekonen, P. Denifl, T. Leinonen, US Patent No. 7820773.
55. A. Pelliconi, C. Cagnani, US Patent Application No. 2006/0047071 A1.
56. J.C. Bailly, R. Hagege, *Polymer*, vol. 32, pp. 181–190, 1991.
57. The catalyst was prepared by the controlled precipitation of dibutyl-magnesium with t-butyl chloride, followed by treatment with ethyl-benzoate and TiCl4. Thus, this is a third generation catalyst.
58. P. Mueller, K.D. Hungenberg, J. Kerth, R. Zolk, US Patent No. 5773537.
59. Z. Zhu, M. Chang, US Patent No. 7700513.
60. F. W. Spaether, D.A. Preskin, US Patent No. 7329626.
61. See discussion in D. Klendworth, F.W. Spaether, US Patent No. 7071137.
62. Single site catalysis is considered as a separate endeavor, separate from traditional Ziegler Natta catalysis.
63. E. Albizzati, P.C. Barbe, L. Noristi, R. Scordamaglia, L. Abrion, U. Giannini, G. Morini, US Patent No. 4971937.
64. G. Morini, E. Albizzati, G. Balbontin, G. Baruzzi, A. Cristofori, US Patent No. 7049377.
65. G. Morini, E. Albizzati, G. Balbontin, G. Baruzzi, A. Cristofori, US Patent No. 7022640.
66. G. Collina, O. Fusco, B. Gaddi, G. Morini, M. Sacchetti, G. Vitale, US Patent App. No 2010/0240846A1.
67. G. Vitale, M. Cimarelli, G. Morini, L. Cabrini, US Patent No. 7033970.
68. M. Mulrooney, T. Merz, 12th Saudi-Japan Symposium, Catalysts in Petroleum Refining & Petrochemicals, KFUPM-Research Institute, Dhahran, Saudi Arabia, 2002.

69. V. Busico, J.C. Chadwick, R. Cipullo, S.Ronca, G. Talarico, *Macromolecules* Vol. 37, 7437–7443, 2004.
70. A. Andoni, Ph.D. Thesis A Flat Model Approach to Ziegler-Natta Olefin Polymerization Catalysts, Eindhoven University of Technology, 2009.
71. K. Blackmon, J.L. Thorman, S.A. Malbari, M. Wallace, US Patent No. 7851578.
72. P. Vincenzi, G. Patroncini, US Patent Appl. No. 2010/0261859A1.
73. G. Patroncini, P. Vincenzi, WO/2009/080497A2.
74. G. Morini, G. Balbontin, G. Vitale, US Patent No. 7202314.
75. G. Balbontin, G. Morini, US Patent No. 7005487.
76. A. Ferraro, I. Camurati, T. Dall'Occo, F. Piemontesi, G. Cecchin, *Kinetics and Catalysis*, Vol. 47, pp. 176–18, 2006.
77. http://www.borealisgroup.com/news-and-events/product-news/2008/borsoft-thin-wall-containers.
78. C.L.Willis, L.H. Slaugh, US Patent No. 4316008.
79. J.M. Kelley, US Patent No. 6586536.
80. J.C. Bailly, D. Durand, P. Mangin, US Patent No. 3933934.
81. G.C. Allen, B.J. Pellon, M.P. Hughes, US Patent No. 4736002.
82. J.M. Kelley, US Patent No. 6998457.
83. L. Sun, G.C. Allen, M.P. Hughes, US Patent No. 5948720.
84. H.G. Becker, H.G. Wey, W. Kilian, T. Stojetzki, M. Vornholt, M. Vey, J. Derks, H.D. Zagefka, US Patent No. 7807768.
85. R. Tharappel, R. Oreins, W. Gauthier, D. Attoe, K. McGovern, M. Messiaen, D. Rauscher, K. Hortmann, M. Daumerie, US Patent No. 6916892.
86. D. Malpass, *Introduction to Industrial Polyethylene*, Scrivener Publishing LLC and John Wiley & Sons, p. 51, 2010.
87. I.V. Eremeev, S.M. Danov, V.R. Sakhipov, and A.G. Skudin, Russian *Journal of Applied Chemistry*, Vol. 74, No. 8, 2001, pp. 141031412
88. T. Iwasaki, T. Jyunke, K. Katayama, K. Tanaka, US Patent No. 6870022.
89. E.S. Shamshoum. D.J. Rauscher, S.A. Malbari, US Patent No. 5330947.
90. http://www.q1.fcen.uba.ar/materias/qi1/Tablas/disocia.pdf
91. Data taken from http://www.polymerpds.akzonobel.com/Polymer ChemicalsPDS/showPDF.aspx?pds_id=462; http://www.akzonobel.com/polymer/our_products/product_search/index.aspx?appid=0&pgid=4&cat=Metal+Alkyls
92. S. Pasynkiewicz, L. Kozerski, B. Grabowski, *J. Organomet. Chem.*, vol. 8, 233–238, 1967.
93. *Polypropylene Handbook*, 2nd edition, Nello Pasquini editor, Hanser Publishers 2005., pp. 35–36 and references therein.
94. R. Spitz, C. Bobichon, M.F. LLauro-Darricades, A. Guyot, L.J. Duranel, *Journal of Molecular Catalysis*, vol. 56, 156. 1989.
95. E. Vahasarja, T.T. Pakkanen, T.A. Pakkanen, E. Iiskola, P. Sormunen, *J. Poly. Sci. Part A: Polym. Chem.*, vol. 25, pp. 3241–3253, 1987.
96. P.C. Barbe, *Adv in Polymer Science*, Springer Verlag, vol. 81, p. 1, 1987.

97. P. Galli, P. C. Berbe, L. Noristi, *Ang. Makromol. Chemie*, vol. 120, p. 73, 1983.

98. K.K. Kang, T. Shiono, Y.T. Jeong, D.H. Lee, *J. Applied Polymer Science*, vol. 71, pp. 293–301, 1999.

99. E.Band, D. Taylor, The Effect of Hydride in Triethylaluminum on Propylene Polymerizations, Akzo Nobel Metal Alkyls Symposium, Scheviningen, The Netherlands, May 1996.

100. http://www.albemarle.com/TDS/Polyolefin_catalysts/sc2010f_ TEA_datasheet.pdf

101. http://www.albemarle.com/TDS/Polyolefin_catalysts/sc2001f_ DEAC_datasheet.pdf

102. Paper char test at 25C.

103. J. Weidlein, *J. Organomet. Chem.*, vol. 32, pp. 181–194, 1971.

104. M. R. Mason, B. Song, K. Kirschbaum, *J. Am. Chem. Soc.* vol. 126, pp. 11812–13, 2004, and references therein.

105. Y.V. Kissin, L.A. Rishina, *Polymer Science Ser. A*, vol. 50, pp. 1101–1121, 2008.

106. A. Alshaiban, Propylene Polymerization Using 4th Generation Ziegler-Natta Catalysts: Polymerization Kinetics and Polymer Microstructural Investigation, Ph.D. Thesis, U. Waterloo, 2011; http://hdl.handle.net/10012/6074

107. A. Dashti, A. Ramazani SA, Y. Hiraoka, S.Y. Kim, T. Taniike, M. Terano, *Polymer International*, vol. 58, pp. 40–45, 2008.

108. M. Abboud, P. Denifl, K.H. Reichert, *J. Applied Polymer Science*, vol. 98, pp. 2191–2200, 2005.

109. M.S. Pimplaupure, X. Zheng, J. Loos, G. Weikert, *Macromolecular Rapid Communications*, vol. 26, pp. 1155–1158, 2005.

110. M. Al-haj, B. Betlem, B. Roffel, G. Weikert, *AIChE Journal*, vol. 52, pp. 1866–1876, 2006.

5

Aluminum Alkyls in Ziegler-Natta Catalysts

5.1 Organometallic Compounds*

Metal alkyls and metallocenes are subsets of organometallic compounds. Interrelationships for a variety of organometallic compounds are depicted schematically in Figure 5.1. The feature that differentiates organometallics is the nature of the carbon-metal bond: metal alkyls are *sigma*-bonded (σ) while metallocenes are *pi*-bonded (π). As shown in Chapter 3, metal alkyls are essential to the performance of Ziegler-Natta catalysts. Metallocenes are used in many single site catalyst systems and are discussed in Chapter 6. Among metal alkyls, the most important for the polypropylene industry are aluminum alkyls. Aluminum alkyls are intimately involved in the mechanism of Ziegler-Natta polymerization of propylene and will be the focus of this chapter.

Most of the commercially available aluminum alkyls are pyrophoric and explosively reactive with water [1–4]. Indeed, inadvertent contact

---

* Portions of this chapter were excerpted from chapter 1 and Appendix chapter A of *Handbook of Transition Metal Polymerization Catalysts*, John Wiley & Sons, Inc., (R. Hoff and R. Mathers, editors), 2010. Used with permission of John Wiley & Sons.

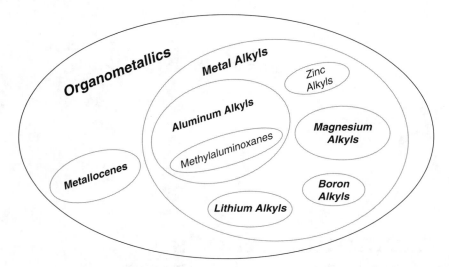

Figure 5.1 Selected types of organometallic compounds. Schematic showing relationships between organometallics, metal alkyls and metallocenes. For industrial polypropylene, aluminum alkyls are, by far, the most important organometallic compounds.

of liquid water with aluminum alkyls in a confined space has been the cause of some of the most severe accidents involving metal alkyls [5]. The resultant explosions are capable of rupturing transfer lines or equipment (tanks, pumps, reactors, etc.) and may cause serious injures to personnel. Hence, it is absolutely essential to purge lines and vessels to insure that no residual liquid water remains before any transfer of metal alkyls is attempted. The final section of this chapter provides guidelines for safety and handling of aluminum alkyls. Considering the hazards of aluminum alkyls, it is remarkable that thousands of kilotons are produced each year and have been supplied for decades to the polyolefins industry worldwide with relatively few safety incidents.

In Ziegler-Natta catalyst systems, aluminum alkyls function in two fundamentally different ways:

- reducing agent for the transition metal compound
- cocatalyst ("activator") for generation of active centers for polymerization

We touched briefly on the function of aluminum alkyls in the mechanism of Ziegler-Natta catalyst systems in Chapter 3. However, the properties of aluminum alkyls and the roles that they play in polymerization will be examined in greater detail below.

Aluminum alkyls are the preferred metal alkyls for Ziegler-Natta catalyst systems for a variety of reasons (see also section 4.10). They perform well with disparate polyolefin catalyst systems and are readily available worldwide. Triethylaluminum (TEAL) is today the most important aluminum alkyl for polypropylene and is sold globally in multi-million pound per year quantities. Other commonly available metal alkyls are either too expensive or perform poorly. For example, R_2Mg compounds deactivate some Ziegler-Natta catalysts when tried as a cocatalyst or "activator." However, magnesium alkyls are highly effective as raw materials for production of supports for Ziegler-Natta catalysts, primarily polyethylene catalysts. The reason for ineffectiveness of magnesium alkyls as cocatalysts may stem from overreduction of the transition metal or blockage of active centers caused by strong coordination of magnesium alkyl.

While in-depth discussions of production of aluminum alkyls are beyond the scope of an introductory text, an overview of key industrial methods will be provided. Detailed discussions of properties, production and applications of aluminum alkyls are available in the literature [6–14]. Cocatalysts for metallocenes will be discussed in the context of single-site catalysts in Chapter 6.

5.2   Characteristics of Aluminum Alkyls

The term "aluminum alkyls" is meant to encompass any compound that contains a carbon-aluminum sigma (σ) bond. Aluminum alkyls include R_3Al, R_2AlCl, $R_3Al_2Cl_3$ (the so-called "sesquichlorides"), $RAlCl_2$, R_2AlOR' and R_2AlH. Though aluminum is nominally trivalent, it is almost always tetracoordinate in aluminum alkyls (see discussion below on association of aluminum alkyls). Among commercially available aluminum alkyls, R is typically a simple C_1 to C_4 alkyl, but is most commonly an ethyl ("Et" or C_2H_5) group.

Methylaluminoxanes are also aluminum alkyls and have become industrially important in recent years as cocatalysts for single site catalysts. However, methylaluminoxanes exhibit significantly different properties than conventional aluminum alkyls, function differently as cocatalysts in olefin polymerization and will be discussed separately in Chapter 6.

As noted earlier, aluminum alkyls are indispensable for industrial polymerization of propylene using Ziegler-Natta catalysts. Principal aluminum alkyls available in the merchant market are provided in Table 5.1. Table 5.1 also provides acronyms commonly

Table 5.1 Principal commercially available aluminum alkyls.

Product*	Acronym	Formula	CAS Number	Theoretical Wt % Al	Comment
trimethylaluminum	TMAL	$(CH_3)_3Al$	75-24-1	37.4	More costly than other R_3Al (see reference 25).
dimethylaluminum chloride	DMAC	$(CH_3)_2AlCl$	118-58-3	29.2	
methylaluminum sesquichloride	MASC	$(CH_3)_3Al_2Cl_3$	12542-85-7	26.3	
triethylaluminum	TEAL	$(C_2H_5)_3Al$	97-93-8	23.6	Most widely used R_3Al for industrial polypropylene
diethylaluminum chloride	DEAC	$(C_2H_5)_2AlCl$	96-10-6	22.4	Used as cocatalyst with early generation PP catalysts.
diethylaluminum iodide	DEAI	$(C_2H_5)_2AlI$	2040-00-8	12.7	Among more costly aluminum alkyls, because of cost of iodine.
ethylaluminum sesquichloride	EASC	$(C_2H_5)_3Al_2Cl_3$	12075-68-2	21.8	Reducing agent in prod'n of 1st generation PP catalysts.
ethylaluminum dichloride	EADC	$C_2H_5AlCl_2$	563-43-9	21.3	
isobutylaluminum dichloride	MONIBAC	$i-C_4H_9AlCl_2$	1888-87-5	17.4	Acronym from "monoisobutylaluminum dichloride."
tri-n-butylaluminum	TNBAL	$(C_4H_9)_3Al$	1116-70-7	13.6	

(Continued)

Table 5.1 (cont.) Principal commercially available aluminum alkyls.

Product*	Acronym	Formula	CAS Number	Theoretical Wt % Al	Comment
triisobutylaluminum	TIBAL	$(i\text{-}C_4H_9)_3Al$	100-99-2	13.6	
diisobutylaluminum hydride	DIBAL-H	$(i\text{-}C_4H_9)_2AlH$	1191-15-7	19.0	
tri-n-hexylaluminum	TNHAL	$(C_6H_{13})_3Al$	1116-73-0	9.6	
tri-n-octylaluminum	TNOAL	$(C_8H_{17})_3Al$	1070-00-4	7.4	
di-n-octylaluminum iodide	DNOAI	$(C_8H_{17})_2AlI$	7585-14-0	7.1	
"isoprenylaluminum"	IPRA**	complex	7024-64-5	14–15%	Viscous, complex composition (see reference 28 of Chapter 5).
diethylaluminum ethoxide	DEAL-E	$(C_2H_5)_2AlOC_2H_5$	1586-92-1	20.7	
ethylpropoxyaluminum chloride	EPAC	(C_2H_5) $(C_3H_7O)AlCl$		17.9	
diisobutylaluminum butylated oxytoluene	DIBAL-BOT	$(i\text{-}C_4H_9)_2AlO(C_6H_2$ $(CH_3)(t\text{-}C_4H_9)_2)$	56252-56-3	7.5	Produced by equimolar reaction of TIBAL (or DIBAL-H) with BHT.***

* As of 2011. Products not necessarily available from all commercial suppliers of aluminum alkyls. (No major supplier offers all of the products listed above.) Aluminoxanes not included (see Chapter 6).

** Also known as ISOPRENYL.

*** BHT also known as 2,6-di-t-butyl-4-methylphenol.

used for these products. Important features and key characteristics of commercially available aluminum alkyls are summarized below:

5.2.1 Basic Physical and Chemical Properties

Most are clear, colorless liquids with low vapor pressures at ambient temperatures. They are miscible in all proportions and compatible with alkanes and aromatic hydrocarbons, but are incompatible with many other common organic compounds (see 5.2.7 below). Most have freezing points well below 0°C (exceptions TMAL, TIBAL and EADC). In neat (solvent-free) form, many ignite spontaneously when exposed to air and are explosively reactive with water. (Please see reference 15 for an in-depth discussion of pyrophoricity of metal alkyls.)

5.2.2 Hydride Content

R_3Al compounds (R = ethyl or higher) contain varying amounts of R_2AlH. Hydride content is expressed in analytical reports as AlH_3 by tacit convention among major suppliers and typically ranges from about 0.02% (wt) in TEAL to about 0.5% in triisobutylaluminum (TIBAL). Expressing hydride content as AlH_3 rather than as R_2AlH is an artifact that makes the R_3Al compound appear to be of marginally higher purity. For example, assume a sample of TEAL has been reported as having the following composition: 94.5% Et_3Al, 5.1% TNBAL and 0.4% AlH_3. If hydride content had been calculated as diethylaluminum hydride, the composition would have been reported as 91.5% Et_3Al, 5.1% TNBAL and 3.4% Et_2AlH.

5.2.3 Other R_3Al Impurities

R_3Al compounds also commonly contain small amounts of other trialkylaluminum compounds (expressed as R'_3Al, where $R' \neq R$). This is usually a consequence of the purity of starting materials or of side reactions during manufacture, such as addition of an ethylaluminum moiety in TEAL across ethylene to produce an n-butylaluminum group as in eq (5.1):

$$\text{AlCH}_2\text{CH}_3 + \text{CH}_2=\text{CH}_2 \longrightarrow \text{AlCH}_2\text{CH}_2\text{CH}_2\text{CH}_3 \qquad (5.1)$$

In most cases, R'$_3$Al contents are low, <0.5% (by wt). An exception is TEAL where n-butylaluminum content from reaction 1 above, expressed as tri-n-butylaluminum (TNBAL) is typically ~5%.

Because rapid alkyl group exchange occurs at ambient T, R'$_3$Al contaminants may be more accurately (and statistically) represented as being present as R$_2$AlR' (sometimes written R$_2$R'Al). However, they are reported analytically as if the extraneous alkyl groups reside on the same aluminum atoms, 3 groups per aluminum. (See also discussion in 5.2.10 below on association of aluminum alkyls.)

As with hydride, expressing extraneous R'$_3$Al compounds in R$_3$Al as R$_2$AlR' would cause purity of the R$_3$Al to be lowered. For example, assume a sample of TEAL has been reported to have the same analysis we used above for the hydride example: 94.5% Et$_3$Al, 5.1% TNBAL and 0.4% AlH$_3$. If TNBAL content had been calculated as Et$_2$Al(n-Bu), the composition would have been reported as 88.7% Et$_3$Al, 10.9% Et$_2$Al(n-Bu) and 0.4% AlH$_3$.

5.2.4   Analysis of Aluminum Alkyls

Industrial R$_3$Al compounds are analyzed by several techniques. A commonly used method is gas chromatographic (GC) analysis of the gaseous products resulting from careful hydrolysis of the aluminum alkyl. For trialkylaluminum compounds, composition is computed by ascribing alkane and hydrogen contents as R$_3$Al and AlH$_3$ (in wt%), respectively, in the unhydrolyzed aluminum alkyl. For RAlX$_2$, R$_3$Al$_2$X$_3$, R$_2$AlX and R$_2$AlOR' compounds, the alkanes from hydrolysis are expressed in mole percent.

Separately, elemental analyses for wt% Al are performed titrimetrically on hydrolysates of the aluminum alkyls. For example, typical aluminum content for TEAL is 23.1% (theory: 23.6%). The observed aluminum content corresponds quite closely to the expected value (*after adjusting for the lowering of expected Al content because of the ~5.0% tri-n-butylaluminum, which theoretically contains only ~13.6% Al*). Assuming a composition of 95% TEAL and 5% tri-n-butylaluminum (and negligible AlH$_3$ content) obtained from the GC analysis, the calculated aluminum content for such a composition would be 23.1%, indicating excellent agreement between observed and expected values. The close agreement between actual and calculated aluminum content indicates that industrial aluminum alkyls have "assays" very near 100%.

In the vast majority of Ziegler-Natta catalyst systems, hydride content and the presence of small amounts of other trialkylaluminum compounds (R'_3Al) are not damaging to performance. However, for selected polypropylene catalysts that employ alkoxysilanes as external donors, hydride can cause a reduction in isotactic content and lowered catalyst activity (16; see also discussion near the end of section 4.10). Tests with TEAL containing up to ~16% of other trialkylaluminum compounds with a modern supported polypropylene catalyst showed no loss of isotacticity and no loss of activity [16].

5.2.5   Impurities Resulting from Exposure to Minute Concentrations of Water and Oxygen

Aluminum alkyls also contain ppm amounts of aluminoxanes and alkoxides resulting from reaction with water and oxygen, respectively (see section 5.6). Water and oxygen enter as minor contaminants (typically <5 ppm) in raw materials used in production of aluminum alkyls. Aluminoxanes and dialkylaluminum alkoxides are usually analytically undetectable (below 500 ppm) and, at these levels, cause no problems in polypropylene catalyst systems.

5.2.6   Assays of Aluminum Alkyls

Total assays are not routinely conducted on commercially available aluminum alkyls. Since impurities mentioned above are also organometallics, total organometallic content of industrial aluminum alkyls typically exceeds 99%. The balance is a combination of inert process oils (a purified white mineral oil is used as lubricant and in agitator seals) and small amounts of solvents (mostly C_6–C_8 aliphatic HC) used to purge reactors and process lines. Observed aluminum contents are very close to expected values, as demonstrated with TEAL above.

5.2.7   Reactivity with Organic Substrates

Aluminum alkyls are highly reactive with many of the common organic solvents. Indeed, reaction with halogenated hydrocarbons (CCl_4, $CHCl_3$, *etc.*) may be explosive after a quiescent period [17]. Organic compounds with labile protons, such as alcohols and carboxylic acids, may also be violently reactive with aluminum alkyls.

Carbonyl compounds, such as ketones, aldehydes and esters, react exothermically with aluminum alkyls. Ethers and tertiary amines also react exothermically to form coordination complexes.

As previously noted, aluminum alkyls are completely miscible and compatible with saturated aliphatic and aromatic hydrocarbons. Large quantities are supplied as 15–25% solutions in aliphatic hydrocarbons such as isopentane, hexane, and heptane. Aluminum alkyls have also been supplied in benzene, toluene and the xylenes, although use of aromatic solvents has diminished in recent years.

5.2.8   Reactivity with CO_2 and CO

R_3Al are reactive with CO_2 [18]. In fact, reaction of TMAL with CO_2 has been used to produce methylaluminoxane cocatalysts for SSC [19–21]. The R_3Al/CO_2 reaction is easily controlled and has been used to passivate aluminum alkyl waste streams [22]. However, R_3Al are unreactive with CO. Also, aluminum alkyls containing halogen or oxygen (DEAC, DEAL-E, etc.) are not reactive with CO_2.

5.2.9   Distillation

As noted above, vapor pressures of aluminum alkyls are typically low. However, low molecular weight aluminum alkyls (C_1, C_2 and $isoC_4$) may be distilled, usually under vacuum which minimizes decomposition. Higher homologs (n-C_4 to n-C_8) are not distillable in industrial process equipment and must be purified by other means, *e.g.*, filtration.

5.2.10   Association of Aluminum Alkyls

As previously mentioned, aluminum alkyls typically involve tetracoordinate aluminum, even though aluminum is nominally trivalent. Most trialkylaluminum compounds are associated as dimers, except when steric bulk of alkyl groups (*tert*-butyl, iso-butyl, etc.) prevents association. Bonding in R_3Al involves three-center-two-electron bonding [23] (also called "electron deficient" bonding [24]), depicted schematically for TMAL in Figure 5.2. At low temperature, proton NMR spectra of TMAL show separate signals for terminal and bridging methyls. At room T, however, rapid alkyl exchange occurs and methyls are indistinguishable by NMR.

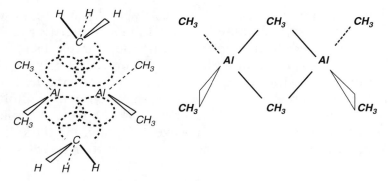

Figure 5.2 Schematic representations of electron-deficient bonding in TMAL dimer. Figure on left depicts partial overlap of sp³ orbitals from bridging methyl groups with lobes of tetracoordinate sp³ aluminums. Figure on right shows more common way of depicting electron deficient bonding in metal alkyls.

Halogenated and oxygenated aluminum alkyls are even more strongly associated, most often as dimers. Heteroatom-containing ligands assume bridging positions. This may be exemplified below by DEAL-E which is dimeric with dative bonds between oxygen and an adjacent aluminum.

5.2.11 Storage Stability

In general, aluminum alkyls are stable indefinitely if stored properly (under dry, inert gas and away from heat). Storage stability of aluminum alkyls may be illustrated anecdotally. At the aluminum alkyls manufacturing site formerly known as Texas Alkyls, Inc. (now Akzo Nobel), a small carbon steel cylinder of DEAC was returned to the Deer Park site after having been stored unopened for about 10 years in a customer's laboratory. The returned product

was sampled and analyzed. The product was still clear (free of particulates), though it had acquired a faint amber tint after 10 years'
storage. Within analytical variance, aluminum and chloride contents and the hydrolysis gas composition were virtually unchanged.

5.2.12   Thermal Stability

Aluminum alkyls demonstrate moderate to excellent thermal stability, depending on ligands. Thermal stability may be important
for aluminum alkyls used in solution processes for polyethylene
since these often operate at >190°C. However, because most polypropylene processes operate below 100°C, thermal stability of the
aluminum alkyl is not as critical. Table 5.2 provides thermal stability
data on selected aluminum alkyls. Lower R_3Als decompose slowly
at elevated temperatures, but thermal stability diminishes as chain
length or branching increases. While quality may be lowered, vio-
lent decomposition does not occur for R_3Al, *except for neat trimethylaluminum. TMAL undergoes potentially hazardous self-accelerating
decomposition above 120°C. Decomposition of TMAL is highly exothermic and is accompanied by generation of large amounts of methane [25].
If confined, an explosion could result.* This potential hazard is moot
relative to PP, because use of TMAL in industrial polypropylene
processes is non-existent at this writing.

Table 5.2 Thermal stability of selected aluminum alkyls*.

Product	T (°C) at Start of Decomposition	% Conversion (3 h @180°C)
triethylaluminum	~120	64
tri-n-butylaluminum	~100	87
triisobutylaluminum	~50	92
tri-n-octylaluminum	60	90
diethylaluminum chloride	174	2
diisobutylaluminum chloride	165	4
diethylaluminum ethoxide	192	0

* G. Sakharovskaya, N. Korneev, N. Smirnov and A. Popov, *J. Gen. Chem. USSR*,
1974, *44*, 560.

For trialkylaluminum compounds other than TMAL, the initial step in thermal decomposition is β-hydride elimination, exemplified in eq (5.2) with TIBAL:

$$isoC_4H_9 — Al \overset{isoC_4H_9}{\underset{\underset{CH_3}{\overset{|}{CH_2— C —CH_3}}}{|}} \overset{H}{\underset{}{}} \quad \xrightarrow{\sim 120°C} \quad (isoC_4H_9)_2AlH + CH_2=C \overset{CH_3}{\underset{CH_3}{}} \qquad (5.2)$$

Further decomposition of the resultant R_2AlH from eq (2) may also occur, resulting in liberation of additional olefin, hydrogen and deposition of elemental Al. However, as previously mentioned, thermal decomposition of aluminum alkyls is essentially a non-factor for polypropylene processes, since most operate at temperatures well below that needed for significant decomposition of aluminum alkyls.

Aluminum alkyls that contain ligands with halogen or oxygen are more stable thermally than the analogous R_3Al compounds. As shown in Table 5.2, TEAL begins to decompose at $\sim 120°C$, while diethylaluminum chloride is stable up to 174°C and diethylaluminum ethoxide up to 192°C.

5.3 Production of Aluminum Alkyls

As of this writing, major manufacturers of aluminum alkyls are:

- Akzo Nobel (formerly Texas Alkyls, Inc.)
- Albemarle (formerly Ethyl Corp.)
- Chemtura (formerly Crompton, Witco and Schering)

These companies supply aluminum alkyls globally and also produce a variety of other organometallics, including metal alkyls of boron, gallium, magnesium and zinc. Akzo Nobel and Albemarle have their principal aluminum alkyl manufacturing facilities in the USA; Chemtura's main site is in Germany. Major manufacturers of metal alkyls have joint ventures and satellite plants around the world. In a few cases, affiliates have manufacturing facilities, but others have only repackaging and solvent blending capabilities.

A few regional suppliers, such as Tosoh Finechem Corporation in Japan, also manufacture aluminum alkyls but have lower capacities and a narrower product range. A joint venture between Albemarle and SABIC was announced in 2009 [26]. The joint venture, to be called Saudi Organometallic Chemicals Company, will have a capacity of about 6000 tons per year of triethylaluminum and is expected to start up in 2012. Sasol recently announced construction of a plant to manufacture triethylaluminum (6000 tons/year capacity) in Brünsbuttel, Germany [27] Capacities for metal alkyls other than TEAL were not disclosed. Also, Petronad Asia (a subsidiary of Kimyagaran Emrooz, an Iranian chemical company) plans to begin production of TEAL in Mahshar in 2013. Petronad plans to supply TEAL to the Middle East market but no capacity figures were released.

Sigma-Aldrich is a US company headquartered in St. Louis, MO and is an excellent supplier of small to medium quantities of organometallic compounds, Sigma-Aldrich provides a wide range of air- and moisture-sensitive organometallic compounds in a variety of innovative packaging. The Sure/Seal package is especially attractive to academia because it permits convenient and safe extraction via syringe techniques of small quantities of these highly reactive materials for laboratory use. However, Aldrich has limited large-scale production capabilities. More information may be obtained at www.sigma-aldrich.com (see especially technical bulletin AL-134).

Many processes are used for manufacture of the wide range of aluminum alkyls (see Table 5.1) offered in the merchant market [4]. Because of the large volumes of R_3Al required by today's polyolefins industry, processes for production of trialkylaluminum compounds are most important. For this reason, an overview of widely used industrial processes for R_3Al production will be provided below.

Karl Ziegler's revolutionary "direct process" was developed in the early 1950s not long after the other exciting discoveries in olefin polymerization made in his laboratories. The first large-scale production of a trialkylaluminum compound *via* Ziegler chemistry was the manufacture of triisobutylaluminum in November of 1959 by Texas Alkyls, Inc. (then a joint venture of Hercules and Stauffer Chemicals, now part of Akzo Nobel). Ziegler's direct process effectively involved reaction of aluminum metal, olefin and hydrogen to produce trialkylaluminum compounds. (This is necessarily an oversimplification of the direct process. Please see references 1–4 and 6–8 for more details). Key reactions involved in the Ziegler direct process for triethylaluminum are shown in eq (5.3) and (5.4):

$$Hydrogenation: 2(C_2H_5)_3Al + Al + 3/2\ H_2 \rightarrow 3(C_2H_5)_2AlH \qquad (5.3)$$

$$Addition: 3C_2H_4 + 3\ (C_2H_5)_2AlH \rightarrow 3(C_2H_5)_3Al \qquad (5.4)$$

Adding equations 5.3 and 5.4 gives the overall reaction for the direct process shown in eq (5.5):

$$Overall\ Reaction: \quad 3C_2H_4 + Al + 3/2\ H_2 \rightarrow (C_2H_5)_3Al \qquad (5.5)$$

However, the reaction shown in eq (5.5) does not take place in the absence of "pre-formed" triethylaluminum.

Up until the early 1990s, large quantities of TEAL were also produced industrially by the so-called "exchange process" which employed the reaction of triisobutylaluminum with ethylene (illustrated in the simplified overall reaction below (eq (5.6)). Isobutylene from eq (5.6) may be recycled. TEAL from the exchange process always contained small amounts of residual triisobutylaluminum (as well as somewhat larger amounts of tri-*n*-butylaluminum from eq (5.1)).

$$(isoC_4H_9)_3Al + 3C_2H_4 \rightarrow (C_2H_5)_3Al + 3isoC_4H_8 \uparrow \qquad (5.6)$$

Both the direct and exchange processes may be run in either batch or continuous mode. Economics favor the continuous, direct process. The direct product is also purer. The exchange process is no longer used for triethylaluminum, but is still used for specialty products such as "isoprenylaluminum" (from reaction of triisobutylaluminum or diisobutylaluminum hydride with isoprene [28]).

The Ziegler direct process technology is vastly superior to historical methods for synthesis of trialkylaluminum compounds. Excellent conversions and yields are obtained with relatively little waste, because all raw materials are incorporated into the product.

Over the past half century, more than 20 aluminum alkyls have been offered in commercial quantities, though some have now become obsolete and are no longer available in bulk. Most of the medium- to high-volume aluminum alkyls in today's market (see Table 5.1) are priced between about $5 and $10 per pound. Exceptions include trimethylaluminum (which is produced by a costly multi-step process [25]) and diethylaluminum iodide (which requires expensive iodine).

Triisobutylaluminum is a commercially available R_3Al that performs comparably to triethylaluminum with many Ziegler-Natta catalysts and typically costs less per pound than TEAL. However, TIBAL contains only about 13.4% Al compared to about 23.1% Al in triethylaluminum. If activity is comparable (*i.e.*, at the same Al/Ti ratio), polyolefin manufacturers buy the product that can provide the greater amount of Al per pound of R_3Al. Because TEAL contains about 70% more aluminum than TIBAL on a molar basis, TIBAL actually costs substantially *more* on a contained aluminum basis. This accounts for the dominance of TEAL in commercial polypropylene processes employing Ziegler-Natta catalysts. Table 5.3 compares the differences in cost of *contained* aluminum of several trialkylaluminum compounds. (Comparison uses hypothetical prices and is for illustration only. For purposes of the comparison, prices (per lb) of TEAL and TIBAL were assumed to be identical, though TIBAL has historically cost less (per lb) than TEAL.)

Aluminum alkyls fulfill several roles in Ziegler-Natta catalyst systems for polypropylene as described in the following sections.

5.4   Reducing Agent for the Transition Metal

This function can be effectively illustrated with a catalyst synthesis used in an early commercial polypropylene process, now obsolete. The catalyst system employed ethylaluminum sesquichloride (EASC) for "prereduction" of $TiCl_4$ in hexane (eq (5.7)). EASC reduces the oxidation state of titanium and $TiCl_3$ precipitates as the β (brown) form which may be converted to the more active γ (gamma) form by heating. Reduction is believed to proceed through an unstable alkylated Ti$+^4$ species (eq (5.7)) which decomposes to Ti$+^3$ (eq (5.8)). Lower oxidation states (Ti$+^2$) may also be formed. These reactions are exothermic and very fast.

Table 5.3 Comparative cost of selected trialkylaluminum compounds.

Product	Assumed Price* ($/lb)	Typical Al Content (wt %)	Cost of Contained Al ($/lb)	Cost of Contained Al (% relative to TEAL)
trimethylaluminum**	50	36.8	136	314
triethylaluminum	10	23.1	43	100
triisobutylaluminum	10	13.4	75	172
tri-*n*-hexylaluminum	10	9.8	102	236

* Hypothetical prices, for illustration only. Contact major manufacturers to obtain actual pricing.

** TMAL is manufactured by a different process than other R_3Al and is much more expensive. See reference 25 of Chapter 5.

$$TiCl_4 + (C_2H_5)_3 Al_2Cl_3 \xrightarrow{\text{hexane}} Cl_3TiC_2H_5 + 2C_2H_5AlCl_2 \qquad (5.7)$$

$$Cl_3TiC_2H_5 \rightarrow TiCl_3 \downarrow + \tfrac{1}{2} C_2H_4 + \tfrac{1}{2} C_2H_6 \qquad (5.8)$$

By-product ethylaluminum dichloride (EADC) from eq (5.7) is soluble in hexane, but is a poor cocatalyst. EADC must be removed (or converted to a more effective cocatalyst) before introduction of monomer. For example, EADC can be easily converted (*via* eq (5.9)) to diethylaluminum chloride by redistribution with triethylaluminum (see reference 29 for discussion of aluminum alkyl redistribution reactions).

$$C_2H_5AlCl_2 + (C_2H_5)_3Al \rightarrow 2(C_2H_5)_2AlCl \qquad (5.9)$$

Aluminum alkyls are still used industrially for prereduction of transition metal compounds. However, far more is used in the role of cocatalyst, described in the next section.

5.5 Alkylating Agent for Creation of Active Centers

In this case, the aluminum alkyl is functioning as a cocatalyst, sometimes called an "activator." Titanium alkyls, believed to be active centers for polymerization, are created through a process known as "alkylation." This occurs when an alkyl is transferred from the cocatalyst to titanium in a ligand exchange reaction. The vast majority of aluminum alkyls sold into the polypropylene industry today is for use as cocatalysts. Molar ratios of cocatalyst to transition metal (Al/Ti) are usually in the range of 40–100 for industrial polypropylene processes using Ziegler-Natta catalysts, but laboratory polymerization tests often employ ratios of 200 or greater.

For "zero," first and second generation polypropylene catalysts (see Chapter 4), DEAC was the cocatalyst that combined acceptable activity and cost with satisfactory polymer stereospecificity. As more advanced, supported catalysts emerged, TEAL provided the best balance of features (cost, activity, stereocontrol, etc.). With

TEAL (today's most widely used cocatalyst), alkylation proceeds as in eq (5.10):

$$(5.10)$$

The titanium alkyl active center may be associated with (or stabilized by) an aluminum alkyl as depicted in eq (3.6) in section 3.7.

5.6   Scavenger of Catalyst Poisons

In commercial polypropylene operations, poisons may enter the process as trace (ppm) contaminants in propylene, comonomer, hydrogen (CTA), nitrogen (used as inert gas), solvents and other process materials. These poisons reduce catalyst activity. Most damaging are oxygen and water. However, CO_2, CO, alcohols, acetylenics, dienes, sulfur-containing compounds and other protic and polar contaminants can also lower catalyst performance. With the exception of CO, aluminum alkyls react with contaminants converting them to alkylaluminum derivatives that are less harmful to catalyst performance. Illustrative reactions of contaminants with TEAL are provided in eq (5.11)–(5.13):

$$(C_2H_5)_3Al + \tfrac{1}{2} O_2 \rightarrow (C_2H_5)_2AlOC_2H_5 \qquad (5.11)$$

$$2(C_2H_5)_3Al + H_2O \rightarrow (C_2H_5)_2Al\text{-}O\text{-}Al(C_2H_5)_2 + 2C_2H_6 \uparrow \qquad (5.12)$$

$$(C_2H_5)_3Al + CO_2 \rightarrow (C_2H_5)_2 AlO\overset{\overset{\displaystyle O}{\|}}{C}C_2H_5 \qquad (5.13)$$

Products from eq (5.11)–(5.13) may undergo further reactions to form other alkylaluminum compounds. Since CO is not reactive with aluminum alkyls, it must be removed by conversion to CO_2 in fixed beds.

As previously noted, Ziegler-Natta catalyst systems used in the polypropylene industry typically employ high ratios of Al to transition metal in the polymerization reactor. Ratios of 40 or greater are common. Hence, there is ample aluminum alkyl available to scavenge poisons and to fulfill the roles discussed in sections 5.4 and 5.5.

5.7   Chain Transfer Agent

Chain transfer for Ziegler-Natta polyolefin catalysts is accomplished largely with hydrogen, as previously shown (see eq (3.7) in Chapter 3). However, at high Al/Ti ratios, molecular weight of the polymer can be marginally lowered by chain transfer to aluminum. This occurs by ligand exchange between titanium and aluminum, illustrated in eq (3.10) of Chapter 3.

5.8   Safety and Handling of Aluminum Alkyls

Industrial polypropylene producers are required to handle aluminum alkyls on a large scale (in some cases, tons per year). As previously noted, many of the commercially available aluminum alkyls are pyrophoric, *i.e.*, they ignite spontaneously upon exposure to air. Hence, all manipulations of aluminum alkyls must be done under an inert atmosphere. (Most often, nitrogen is used.) Furthermore, most commercially available aluminum alkyls are explosively reactive with water. Despite an abundance of resources and training aids from aluminum alkyl suppliers, accidents occur and severe injuries and even death have resulted. Clearly, safety and handling of aluminum alkyls must be a high priority.

Safety measures that should be taken while handling aluminum alkyls are best viewed as a cascading series of defenses against mishap. The first line of defense must be the person at the "front line", *i.e.*, the person actually making the transfer. That person must be thoroughly trained and practiced on safe procedures for transferring aluminum alkyls. He or she must not take short cuts or circumvent precautions.

The second line of defense is to insure that equipment to be used in the transfer is well-designed and properly assembled. Transfer lines must include purge-capability following transfer. In preparation for a transfer, the equipment must be thoroughly checked using inert gas pressure to insure that all connections are leak-free (soap solution is typically used to detect leaks). As previously indicated, it is absolutely imperative to insure that no free-standing water is in transfer lines or equipment. Failure to remove water can result in explosions that can cause severe injuries or death. Further, all transfer lines should be dedicated to aluminum alkyls service, *i.e.*, there should be no extraneous connections to the transfer line that could contain water or other potentially reactive chemicals.

While handling aluminum alkyls, the last line of defense is to insure that the appropriate personal protective equipment (PPE) is worn. Major commercial suppliers recommend a variety of personal protective equipment for large-scale handling of aluminum alkyls including a flame-resistant hood, fire resistant coat and leggings, and impervious gloves (leather gloves or felt-lined PVC are most often used). Technical bulletins covering procedures for large-scale handling and transfer of aluminum alkyls are available from the major producers listed in section 5.3.

For small-scale laboratory transfers, recommended personal protective equipment includes a *minimum* of a full face shield and fire retardant lab coat. A detailed discussion of techniques for safe laboratory transfers of air-sensitive compounds is available (30).

Irrespective of the quantity of aluminum alkyl to be transferred, the worker wearing proper personal protective equipment will likely avoid severe injury in the event of an accident resulting from an unsafe act or equipment failure.

5.9   Questions

1. What are organometallic compounds? Why are they important in the polypropylene industry?
2. You have received a shipment of TEAL with the following reported analysis: 94.1% TEAL, 5.3% TNBAL and 0.6% AlH_3. What would the analytical values have been had hydride content been expressed as diethylaluminum hydride?
3. What are the key functions of aluminum alkyls in ZN polymerization of propylene? Illustrate each role with equations.
4. Why are DEAC and TIBAL typically not used as cocatalysts with supported ZN catalysts?

5. TEAL is today the most widely used trialkylaluminum cocatalyst for PP but may contain small amounts of contaminants. What is the most damaging impurity in TEAL to ZN polypropylene catalysts? What is the effect of other R_3Al in TEAL?

6. Why is thermal stability of aluminum alkyls not a significant concern in polypropylene ZN catalyst systems?

7. A polypropylene plant has received a shipment of TEAL with the following analysis: 95.0% TEAL, 5.0% TNBAL, <0.01% AlH_3 and an aluminum content of 22.5%. Though these values are within specifications for TEAL, they suggest a problem with the product. What is the problem?

References

1. JR Zietz, Jr., GC Robinson and KL Lindsay, *Comprehensive Organometallic Chemistry*, Vol 7, 368, 1982.

2. JR Zietz, Jr., *Ullman's Encyclopedia of Industrial Chemistry*, Vol. A1, VCH Verlagschellshaft, Weinheim, FRG, 543, 1985.

3. MJ Krause, F Orlandi, AT Saurage and JR Zietz, Jr, *Ullman's Encyclopedia of Industrial Chemistry*, Wiley-VCH Verlag GmbH & Co. KGaA, Weinheim, 2005.

4. DB Malpass, LW Fannin and JJ Ligi, *Kirk-Othmer Encyclopedia of Chemical Technology*, John Wiley and Sons, New York, Third Edition, Volume 16, 559, 1981.

5. O Rentas, *Akzo Nobel Metal Alkyls Symposium 1996*, Scheveningen, The Netherlands, May, 1996.

6. JR Zietz, Jr., *Ullman's Encyclopedia of Industrial Chemistry*, Vol. A1, VCH Verlagschellshaft, Weinheim, FRG, 1985, p 543.

7. JJ Ligi and DB Malpass, *Encyclopedia of Chemical Processing and Design*, Marcel Dekker, New York, Vol 3, 1, 1977.

8. F Bickelhaupt and O Akkerman, *Ullman's Encyclopedia of Industrial Chemistry*, Vol. A15, VCH Verlagschellshaft, Weinheim, FRG, 626, 1985.

9. DB Malpass, *Handbook of Transition Metal Polymerization Catalysts*, R Hoff and R Mathers (editors), Wiley, 1, 2010.

10. FR Hartley and S Patai, *The Chemistry of the Metal Carbon Bond*, John Wiley and Sons, New York: Vol 1, *The Structure, Preparation, Thermochemistry and Characterization of Organometallic Compounds*, 1983; Vol 2, *The Nature and Cleavage of Metal-Carbon Bonds*, 1984; Vol 3, *Carbon-Carbon Bond Formation Using Organometallic Compounds*, 1985; Vol 4, *The Use of Organometallic Compounds in Organic Synthesis*, 1987.

11. JJ Eisch, *Comprehensive Organometallic Chemistry*, Vol 1, p 555, 1982.

12. JJ Eisch, *Comprehensive Organometallic Chemistry II*, Vol 1, p 431, 1995.

13. WE Lindsell, *Comprehensive Organometallic Chemistry*, Vol 1, p 155, 1982.

14. WE Lindsell, *Comprehensive Organometallic Chemistry II*, Vol 1, p 57, 1995.

15. DB Malpass, *Handbook of Transition Metal Polymerization Catalysts*, R Hoff and R Mathers (editors), Wiley, 551, 2010.

16. E Band and D Taylor, *Akzo Nobel Metal Alkyls Symposium 1996*, Scheveningen, The Netherlands, May, 1996.

17. Though a few combinations of halogenated hydrocarbons with aluminum alkyls are stable, others may decompose violently. In some cases, solutions give the initial appearance of compatibility, but may decompose explosively after an induction period. Extreme caution is urged. See WH Thomas, *Ind. Eng. Chem. Prod. Res. Dev.*, *21*,120, 1982.

18. Reactivity of R_3Al compounds with CO_2 has been known for decades (see K Ziegler, *Organometallic Chemistry*, (ACS Monograph 147, H. Zeiss, editor), Reinhold, NY, 240, 1960), but it was not until the 1990s that methylaluminoxanes from $TMAL/CO_2$ reaction were shown to have utility as cocatalysts for SSC. See references 19–21.

19. GM Smith, JS Rogers and DB Malpass, *Proceedings of the 5th International Congress on Metallocene Polymers*, Düsseldorf, Germany, organized by Schotland Business Research, Inc., Skilman, NJ, March 31-April 1, 1998.

20. GM Smith, JS Rogers and DB Malpass, *Proceedings of MetCon '98*, organized by The Catalyst Group, Spring House, PA, June 10–11, 1998.

21. GM Smith, SW Palmaka, JS Rogers and DB Malpass, US Patent 5,381,109, Nov. 3, 1998.

22. AM Piotrowski and JJ Ligi, US Patent 4,875,941, October 24, 1989.

23. JP Collman, LS Hegebus, JR Norton and RG Finke, *Principles and Applications of Organotransition Metal Chemistry*, University Science Books, Sausalito, CA, 100, 1987.

24. K Ziegler, *Organometallic Chemistry*, (ACS Monograph 147, H. Zeiss, editor), Reinhold, NY, 207, 1960.

25. D.B. Malpass, *Methylaluminum Compounds*, Society of Plastics Engineers (SPE) *The International Polyolefins Conference*, Houston, TX, February 25–28, 2001

26. AH Tullo, *Chemical & Engineering News*, 14, November 2, 2009.

27. AH Tullo, *Chemical & Engineering News*, 16, June 21, 2010.

28. JJ Ligi and DB Malpass, *Encyclopedia of Chemical Processing and Design*, Marcel Dekker, New York, Vol 3, 32, 1977.

29. T. Mole and E Jeffery, *Organoaluminum Compounds*, Elsevier, New York, 30, 1972.

6

Single Site Catalysts and Cocatalysts

6.1 Introduction

The common chemical limitation of heterogeneous catalysts, including Ziegler-Natta catalysts, is that they are complex mixtures of components, containing a variety of catalyst site structures at their surfaces. They are prepared more by recipe than by synthesis. It is inevitable, then, that chemists have long been studying discrete transition metal compounds as a more systematic approach to olefin polymerization catalysis. This subject has been intensively investigated as a counterpoint to Ziegler-Natta catalysis since the 1950s, and pioneering work by several laboratories revealed that ethylene polymerized rapidly in combination with bis(cyclopentadienyl) titanium dichloride and alkylaluminums in the presence of water [1, 2, 3]. The homogeneous nature of the polymerization system was conducive to detailed study, particularly by NMR characterization. Though interesting from an academic viewpoint, these early versions of single site catalysts were very low in activity and unsuitable for industrial applications.

However, two other factors are more significant motivations for study of molecular polymerization catalysts. The first is that

molecularly discrete catalysts lead to uniform polymer compositions, because all catalyst sites are essentially the same. Polymer
uniformity encompasses factors related to chain length, statistical
comonomer incorporation, and random insertion errors. This leads
to new physical properties and new copolymer types, not achievable
in the less uniform polymers from ZN catalysts. Secondly, systematic
and well-characterized modifications of the catalyst structure are
possible, leading to understanding and control of the extent of stereo
defects and regio defects in the polypropylene. This leads directly to
the development of new polymers with controlled properties.

Accordingly, a very large body of literature now documents the
design and synthesis of discrete transition metal complex catalysts that permit the precise control over polymer microstructure.
Structurally, most work has been on a class of organometallics
called metallocenes, and the derived resins are referred to as metallocene polypropylenes, or mPP. The features of mPP are discussed
in section 6.4, with the most salient feature being narrow MWD.
This is a direct consequence of the homogeneity of the catalyst
structure, and differentiates mPP resins from polypropylene made
with Ziegler-Natta catalysts.

Patent activity has been consistent, with over 200 new patents
per year for the last fifteen years, mostly from the polypropylene-
producing industry. Collectively, the metallocene catalysts and
other discrete catalyst types are called single site catalysts, SSCs.
Several excellent comprehensive reviews are available to the interested reader [4].

6.2 The Structures of Metallocenes and SSCs

Tacticity control is the primary target in defining the structures of
metallocenes for polypropylene. Control of tacticity derives from
one of two mechanisms: control by the chirality of the catalytic center, or control by the chirality derived from the last inserted monomer. The former approach is more robust, so, as with ZN catalysts,
the focus is to control PP tacticity *via* the stereochemical environment of the metallocene catalyst.

Metallocene compounds consist of a transition metal bound to
one or two (substituted) cyclopentadienyl (Cp) ligands as well as
ancillary ligands [5]. The Cp binds to the metal *via* an aromatic π
system in a penta-hapto ($"\eta^5"$) coordination mode [6]. For unsubsti-
tuted Cp there is usually free rotation of the Cp ring about the ring

centroid-to-metal axis. However, when two Cp ligands coordinate the metal and are joined to each other by a one- or two-atom bridge, then a relatively rigid or constrained structure results. Members of this latter class are called *ansa*-metallocenes, and, depending upon the structure, they may be chiral, and the chirality is focused at the metal center. *Ansa*-metallocenes are useful for creating mPP resins. The nature of the bridge and the substitution pattern on the Cp ligands are crucial in determining the nature of the catalyst activity and polymer tacticity. Substitution on the Cp often involves a second, fused ring.

Most industrially useful *ansa*-metallocenes for production of polypropylene are from the Group 4 metals: Ti, Zr, and Hf. The latter elements comprise the most useful catalysts for mPP and are in the formal +4 oxidation state (d^0), usually as a cationic (or partially cationic) species, paired to a bulky, weakly coordinating anion. Two labile ligands are required for attachment at the metal center: one to be replaced by an alkylating agent or the growing polymer chain, and the other to be replaced by incoming monomer. That is, the metal complex must contain at least two coordination sites which can be activated: one to provide a metal-alkyl bond and one to provide a coordination site for the incoming olefin. It is usually envisioned that the sites are cis to each other, as shown in Figure 6.1, to facilitate the olefin insertion step.

The significance of metallocenes, or of any discrete catalytic compound, is that the scientist can correlate the established catalyst structure with performance, hypothesize a mechanism, and make rational changes to the structure to modify that performance. For *ansa*-metallocenes the main design parameters are:

1. the symmetry about the metal center,
2. the steric congestion about the metal center, and
3. the rigidity of the ligand structure.

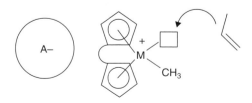

Figure 6.1 Schematic representation of an activated *ansa*-metallocene. A- represents the bulky anion, and □ represents a vacant coordination site. After insertion of the monomer the resulting alkyl group becomes the growing polymer chain.

Much elegant work has been done to correlate these three factors with the microstructure of the produced mPP, and this has permitted a rather rapid development of structures, tailored for specific resin targets [7].

Metallocene symmetry is the basis for achieving steric control during polymerization. Metallocenes with high symmetry, such as two planes of symmetry, catalyze formation of atactic polymer [8]. Metallocenes with one mirror plane (C_s symmetry) are useful for producing syndiotactic polypropylene (see section 6.9). Metallocenes lacking a mirror plane of symmetry are useful for producing iPP. The latter structures are typically members of the C_2 and C_1 symmetry groups [9]. How does the symmetry arise? It is the result of the architecture of the ligands around the metal center and the rigidity of that structure. Rigidity is achieved by a bridging atom or atoms, called a linker, which locks the configuration about the transition metal, restricting rotation of the ligands, as shown in Figure 6.2.

By varying the Cp substitution (including fused rings) and the nature of the linker, synthetic chemists have refined the rigidity of the structures and steric demands for propylene complexation and for the growing polymer chain. This has led to diverse, structurally sophisticated families of molecules tailored to produce various PP microstructures. A sampling is shown in Figure 6.3 to convey the breadth and complexity of the structures appearing in the US patent literature, with an emphasis on recent literature. It is not intended to define which molecules have the most commercial potential.

The structures in Figure 6.3 demonstrate several points. Zr and Hf are the transition metals of choice, in contrast to Ziegler Natta catalysts

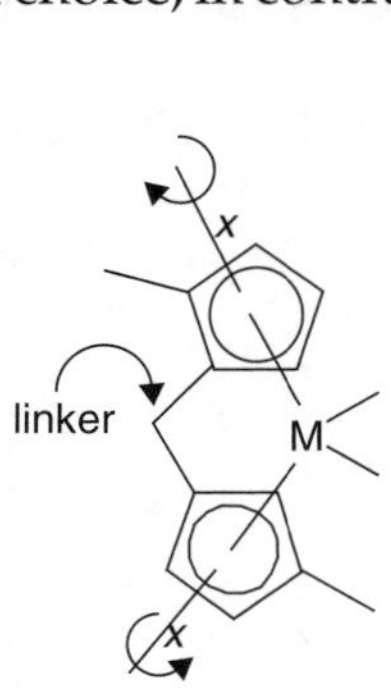

Figure 6.2 Schematic representation of restricted rotation of an *ansa*-metallocene. The rotation of the substituted Cp ligands around the Cp-M axis is prevented by the linker atom which "locks" the configuration around the metal center.

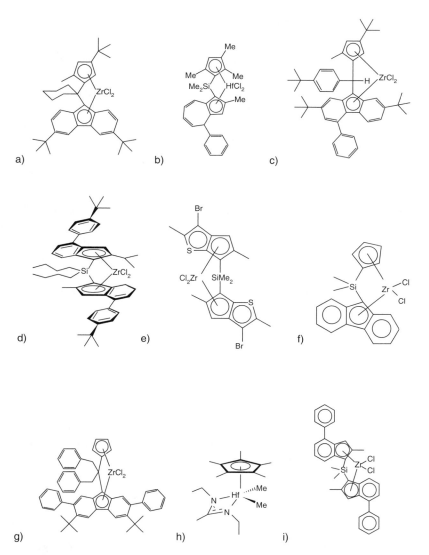

Figure 6.3 Some recent metallocene structures for polypropylene. a) Reproduced from US Patent No. 7449533, [10] iPP b) Reproduced from US Patent No.7906599, [25] iPP c) Reproduced from US Patent No.7470759, [11] iPP d) Reproduced from US Patent No. 7285608, [12] iPP e) Reproduced from US Patent No. 7868197, [13] iPP f) US Patent No. 4892851, [58] sPP g) Reproduced from US Patent Application No. 2010/0069588, [14] sPP h) Reproduced from US Patent Application No. 20110028654, [15] aPP i) US Patent No. 6051727, [16] iPP.

for PP which are based on Ti. The aromatic penta-hapto rings include substituted cyclopentadienyl, indenyl, and fluorenyl types. Systems have also been developed using only one Cp ring, the so-called "half-sandwich" metallocenes. The aryl groups can be substituted with heteroatoms. Bridging atoms are both carbon and silicon and can include bulky substituents. In those structures with a mirror plane, atactic or syndiotactic PP may result, depending upon the structure. Thus, a wide array of structures have been prepared and evaluated.

Catalytic activity of SSCs can be very high, approaching 1,000,000 g iPP/mmol M.h [17] where M is Zr or Hf. By comparison a supported ZN catalyst activity is about 100,000 g iPP/mmol M.h, where M is Ti. The difference may simply be due to the small fraction of total Ti centers that are catalytically active in the ZN catalyst. Once immobilized on a support, SSC activity is similar to, or even below, supported ZN catalysts on a per weight catalyst basis [18]. This is an economic consideration.

Recall that a ZN catalyst is a diverse cocktail of catalyst sites. Therefore ZN catalysts produce, a cocktail of polymer chain structures. SSCs do not have this characteristic. They tend to produce more uniform polymer chain structures in the sense that all the polymer chains resemble each other. Similarly, stereo-imperfections and the extent of comonomer incorporation are also relatively uniformly (i.e., randomly) distributed across chains. Thus, whereas the art of ZN catalyst design relies upon modifying the distribution of the catalyst cocktail sites to control resin properties, the design of metallocene catalysts relies upon a specific structural change that affects all catalyst sites evenly. If one might risk an analogy, ZN catalysts are analog technology, whereas SSCs are products of digital technology. Both technologies have a place in industry.

6.3 Non-Metallocene Polymerization Catalysts

Brief mention is made here that there are large numbers of discrete polymerization catalysts, and not all are metallocenes. Some catalysts have been developed based on the Group 4 metals with ligand structures that do not rely upon the η^5 cyclopentadienide skeletal moiety. For example, chelating complexes based on structurally substituted variations of the Salen ligand or tridentate nitrogen chelates have been reported to catalyze production of isotactic polypropylene [19].

Salen ligand

Others have been developed outside of the Group 4 metals, with particular focus on the late transition metals (Groups 9 and 10), such as Ni and Co. However, these catalysts are typically used for ethylene polymers and copolymers and seldom employed for producing polypropylene. Perhaps the most significant feature of the late transition metal complexes is their tolerance of polar monomers, permitting the preparation of polymers inaccessible from Ti or Zr catalysts, which are too highly oxophilic. Reviews are available for the interested reader [2, 20]. These catalysts are not believed to be in commercial use at the time of this writing.

6.4   Cocatalysts for SSCs

The cocatalyst is just as essential for SSCs as for ZN catalysts. For supported ZN catalysts the cocatalyst functions to reduce and alkylate the titanium center. For SSCs the cocatalyst functions to form a cation and alkylate the metal center. This section describes two historically important types of SSC cocatalysts: oligomeric aluminoxanes and organoboron compounds (aka "molecular cocatalysts"). The latter are often employed in combination with a small amount of an aluminum alkyl. This section also briefly describes the recent development of activated supports which promote formation of the cation and that may impact upon aluminoxane and molecular cocatalyst usage.

6.4.1   Aluminoxanes

Historically, the preferred activator for SSCs has been methylaluminoxane, MAO, and it still predominates in the literature today. MAO is a mixture of oligomers produced by the carefully controlled partial hydrolysis of trimethylaluminum (TMAL) with water in an inert hydrocarbon [1, 21]. MAO is a complex mixture comprising multiple equilbria between different structures. It is believed to contain

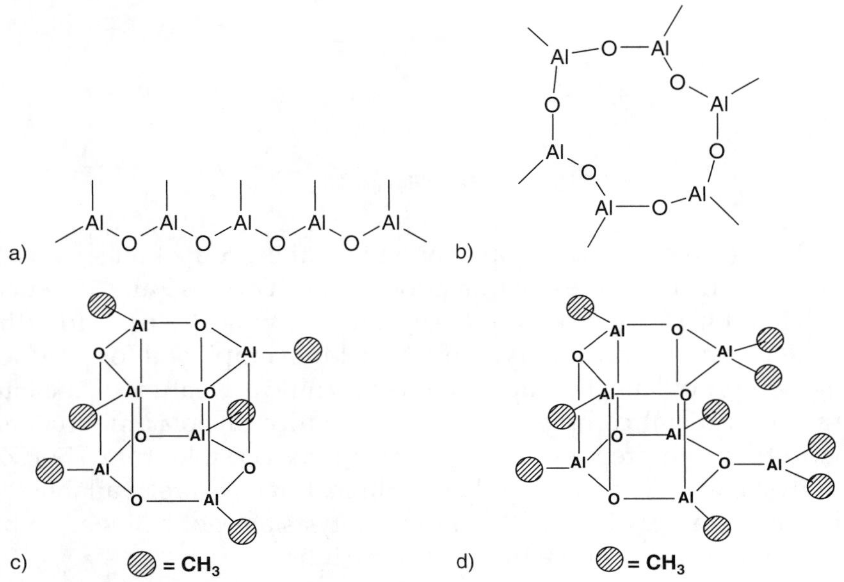

Figure 6.4 Structural types in methylaluminoxane. a) chain structure; b) ring structure; c) Cage structure with hexagonal and square faces d) cage structure with two square faces opened by reaction with TMAL.

chain, ring, and cage architectures (Figure 6.4). Three dimensional structures with hexagonal and square faces (and sometimes octagonal faces) have been calculated to be the most stable [22]. MAO also contains residual free TMAL, and structures have been proposed in which the TMAL is coordinated to the MAO skeleton, involving tetracoordination at aluminum centers. For example, Barron has proposed a nonameric cluster structure for MAO [23] based upon work with *tert*-butylaluminoxane. Moreover, stripping the TMAL can generate additional TMAL, which speaks to the dynamic structural situation. In practice, the structures are a function of the exact extent of hydrolysis and may also be a function of how the hydrolysis was conducted.

The MAO functions to methylate the transition metal by ligand exchange, and then abstracts a methide (CH_3^-) to form a cationic metal center. Together with the aluminate, they comprise a weakly coordinated ion pair. This constitutes the active catalyst for polymerization and is shown schematically in Figure 6.5:

Cocatalyst cost is an important economic factor. MAO cocatalysts are normally required to be used in large excess (50–1000 Al/M)

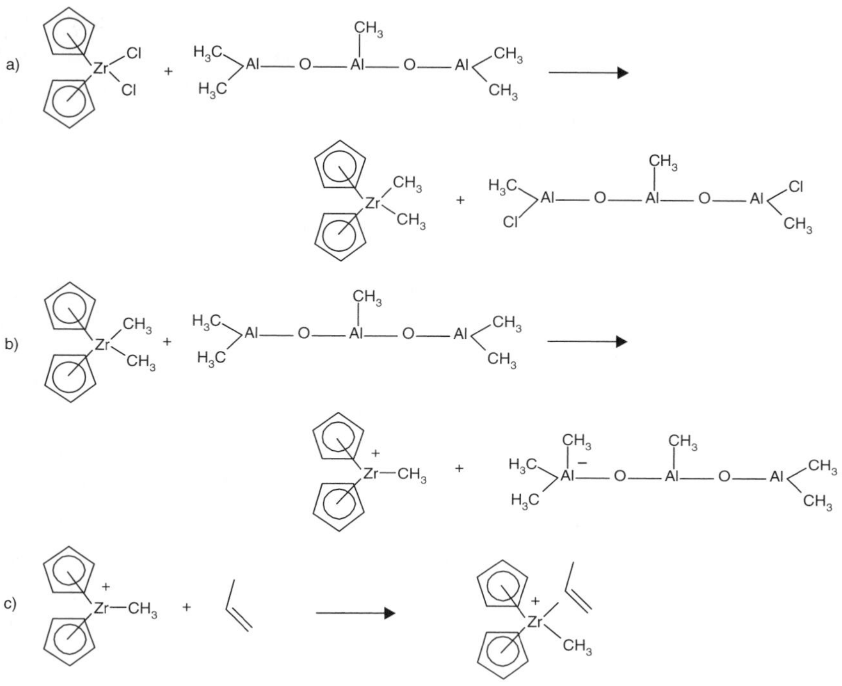

Figure 6.5 Simple view of metallocene activation by excess MAO [24].
a) methylation b) formation of ion pair c) coordination of propylene [25].

compared to the transition metal in order to achieve high activities. This may be due to MAO's structural complexity, where only a fraction of the structures accomplish ion pair formation, or it may be due to a small ion pair formation constant, requiring a large excess of MAO to drive the equilibrium to the ion pair. Additionally, the cost of MAO is documented to be much higher than the organoaluminum compounds used with ZN catalysts [26]. Partial replacement of MAO with conventional aluminum alkyls and new types of MAO continue to be developed in the patent literature, with a goal to reduce cost. There are also some new versions which claim higher activities [27, 28, 29]. Alternative versions include the so-called "modified methylaluminoxanes" (MMAO) and methylaluminoxanes produced by non-hydrolytic methods. However, these alternatives are not as broadly useful with metallocene SSC as "conventional" MAO and should be regarded as niche products [18].

It is interesting to note that the metallocene-aluminoxane system is, in a sense, an inverse image of the Z-N catalyst system; the former consist of a structurally discrete transition metal component and a structurally complex cocatalyst, whereas the Z-N catalyst is a structurally complex transition metal component with a structurally discrete cocatalyst. Newer developments are described in sections 6.4.2 and 6.4.3.

6.4.2   Organoboron Cocatalysts

In addition to MMAOs and non-hydrolytic versions mentioned above, alternative activators called "organoboron cocatalysts" or "molecular cocatalysts" have been developed. The complicated manufacture of MAO, its cost, and its poorly defined structure led to the investigation of discrete compounds capable of forming weakly held ion pairs with the metallocene instead of MAO. The principal molecules that have been reported as useful thus far include tris(pentafluorophenyl)boron (FAB), dimethylanilinium tetrakis(pentafluorophenyl)borate (DAN-FABA), and triphenylcarbonium tetrakis(pentafluorophenyl)borate (TRI-FABA) [30]. The nominal reaction for activation of DAN-FABA is shown in eq. (6.1), below, with the active ion pair in bold. Dimethylaniline and methane are byproducts from the activation reaction. While the latter is clearly inert, the former nominally could play some role in polymerizations as a Lewis base.

$$Cp_2M(CH_3)_2 + HN(CH_3)_2C_6H_5\ B(C_6F_5)_4 \rightarrow$$
$$\mathbf{[Cp_2M(CH_3)]^+} + \mathbf{[B(C_6F_5)_4]^-} + N(CH_3)_2C_6H_5 + CH_4 \qquad (6.1)$$

where M = metal, typically Ti, Zr, Hf

Fluorinated boron compounds are used in near stoichiometric combination with the metallocene to form ion pairs. They are normally employed with TEAL or some other trialkylaluminum, which acts as a poison scavenger and realkylating agent. Other, more complex fluorinated ligands have been developed and shown to function even in the absence of aluminum [31]. It is unclear if the cost of these high molecular weight specialty organics provides a clear performance or cost benefit over MAO.

6.4.3   Activated Supports

An area of recent development is novel activated supports [32]. While the general role of supports is active center immobilization, discussed in Section 6.5, activated supports have a dual role of both metallocene supportation and metallocene activation, and they are (somewhat arbitrarily) considered here within Section 6.4. These supports are prepared by treatment of aluminas, silicas, alumina-silicas or other solid oxides with fluorinating agents (examples: NH_4F, AlF_3), sulfating agents (example: $(NH_4)_2SO_4$), or with chlorinating agents (examples: metal halides, CCl_4). This is typically combined with a calcination step which reduces the surface hydroxyl population. The additive to the oxide matrix is considered to be an electron withdrawing source, which increases acidity of the oxide support [33]. The treated support acts to promote the formation, or partial formation, of the cationic metal center by eq. (6.2), which is a reaction similar to eq. (6.1) above: [34, 35]

$$(6.2)$$

sulfated alumina surface   sulfated alumina surface
                                            with delocalized negative charge

This reduces or eliminates the need for MAO or FAB derivatives. If the metallocene already contains an alkyl group, then an aluminum alkylating agent may not be needed either. Most reports, however, show that a conventional aluminum alkyl like TIBAL is used as an alkylating and poison scavenging agent. Among non-metallocene, chelated catalysts (section 6.3), it has been found that $MgCl_2/R_nAl(OR')_{3-n}$ is a good activator and support for polymerization in the absence of MAO [36]. Thus, activated supports may hold promise to supplant MAO and FAB cocatalyst activators in the future.

6.5 Supports for SSCs

In ZN catalysts the support and ancillary donors establish the chirality about the Ti center and may serve to improve the activity of the Ti center. However, metallocenes have their chirality synthetically engineered into the transition metal complex. A support is still very important in order to prevent the polymerization reaction from producing a difficult-to-handle, low bulk density powder, which coats reactor surfaces and impedes heat reactor exchange. The essential support role for a SSC is to impart the needed polymer morphology, which is accomplished by anchoring the active site to the support surface and relying upon the replication phenomenon described in Chapter 3. However, note the development of activated supports in Section 6.4.3 above which expands the support role.

Many supports are suitable for immobilization, provided that they have the right characteristics. Silica is the most common support, and SSCs may use commercial grades already employed for ZN ethylene polymerization catalysts. Other supports described in the literature include alumina, zeolites, clays, polystyrene, polyolefins, polyamides, talc, MgO, and ion exchange resins [37, 38, 39, 40]. Key factors in support performance include surface area, pore size, pore volume, particle size distribution, mechanical strength, and surface hydroxyl population. Higher pore volumes generally mean higher catalyst activity. It has been found, fortuitously, that immobilizing the metallocene results in reducing the required amount of cocatalyst, and, in some cases, permits the use of TEAL in place of aluminoxanes [41, 42].

There are nearly a dozen methods reported in the literature for anchoring the metallocene and activator to the support surface, and the reader is referred to reviews on the subject [43]. The goals of the supportation method include immobilization of the metallocene, distribution uniformly throughout the support, chemical activation, preservation of the metallocene structural integrity, and an industrially practical process. The methods reported cover a broad range from simple physical deposition by precipitation, to covalent tethering of metallocenes by spacer group structures [44]. The methods will depend strongly upon the activator type: MAO (section 6.4.1), FAB (section 6.4.2.), or an activated support (section 6.4.3.) In the context of MAO, according to Fink *et al.* [37] the best method is the one step immobilization of a preactivated metallocene/MAO complex upon a silica support with a defined hydroxyl content. Hydroxyl content is controlled by precalcination of the support. A representative sequence of the aforementioned supportation is given below:

a) Calcine the silica to a predetermined hydroxyl concentration.
b) Dissolve the metallocene in a toluene solution of MAO. The Al/metallocene may be in the ~50–1000/1 range.
c) Slurry the silica in the metallocene/MAO solution and mix.
d) Filter, and wash the solids with toluene to remove soluble materials.

6.6   Characteristics of mPP

As mentioned previously, the most salient feature of mPP is its narrow MWD. This is a direct consequence of the homogeneity of the catalyst structure, and differentiates mPP resins from ZN PP resins. A typical mPP has a polydispersity index, M_w/M_n ~2–3, whereas for a ZN PP it ranges from ~4–8. The lack of comparatively long and short chain lengths vs. the M_w impacts various properties. For example, odor and taste are generally associated with smaller molecules, as is smoke generation during certain plastic processing. Short chains act as lubricants during melt processing. Long chains are associated with higher viscosity and melt strength, which also impacts processing characteristics.

The second distinctive property of mPP is its lower melting point compared to ZN iPP. This is traced to the occurrence of occasional 2,1 insertion defects characteristic of many metallocenes catalysts. These defects are largely absent in ZN PP. The 2,1 insertion creates crystalline imperfections which lower the melting point of otherwise isotactic structures. By contrast, even with a sizeable xylene soluble (XS) fraction, the melting point of ZN iPP is ~160°C. Some data demonstrating this is shown in Table 6.1 [45]. The ZN iPP has

Table 6.1 Occurrence of 2,1 insertions in polypropylene.

Property	ZN iPP Resins	Metallocene iPP Resins
Xylene Solubles, %	0.7–4.5	0.2–0.3
mmmm, %	95–97	94–96
Racemic, %	0.2–1.5	0.7–1.5
2,1 insertion, %	0	0.6–1.0
Tm, °C	160–163	150–151

a significantly higher melting point despite higher XS. Tacticity, in terms of meso and racemic analysis are roughly comparable. Thus, 2,1 insertion may be the root cause of the typically lower melting point of metallocene iPP. The lower crystallinity leads to smaller crystalline domains and better transparency in films, without production of excessive soluble polymer. Systematic adjustments of the molecular structure can control the extent of structural defects and the effect upon physical properties. For example, as the rr triads increase (which are due to a single insertion error - see section 2.2.2 and Figure 2.4), the crystallinity, melting point and Young's modulus decrease, while ductility and polymer toughness increase [46, 47, 48].

Comonomer incorporation is more uniform in mPP than in ZN PP. Consequently, it is possible to add higher levels of ethylene and other alpha olefins without generating the soluble polymer that would be a result of short chain blocky copolymers. Very high comonomer levels have been achieved, giving rise to resins with very high impact resistance, similar to EPDM rubbers [49]. These are termed metallocene plastomers and have been commercialized by several polymer manufacturers.

Metallocene catalysts usually produce a relatively high amount of chain end unsaturation, such as from beta hydride elimination, unlike ZN catalysts. The terminal double bond affords the possibility of polymer functionalization, such as by free radical, ene or Diels Alder reactions to add maleic anhydride [50]. Functionalizing PP improves adhesion properties, such as paintability (see discussion in section 11.3).

A brief listing of mPP properties is given in Table 6.2, summarizing some important resin qualities.

Table 6.2 Some characteristics of mPP resins.

Resin Property	Characteristics and Comments
Comonomer incorporation	High levels are achievable with a more random distribution than for ZN resins. However, in many cases ethylene incorporation leads to low MW [51]
Gas barrier properties	Low gas permeability of high MW mPP resins is due to the absence of lower, more mobile chain lengths

(Continued)

Table 6.2 (cont.) Some characteristics of mPP resins.

Resin Property	Characteristics and Comments
Impact resistance	Reducing melting point generally reduces resin brittleness without strongly increasing xylene solubles. Impact resistance improves and polymer stiffness is largely retained. Very high comonomer incorporation possible, forming plastomers [39]
Low molecular weight resins	Low MW is readily achievable and good for specialty waxes.
Melt viscosity	Narrow MWD typically lowers melt strength vs ZN resins, which is a disadvantage in some forming operations. However, new mPP grades are being introduced to improve melt strength [52] Low melt viscosity is an advantage for nonwoven fibers.
Melting point	Typical T_m is ~145–150°C, which is below homopolymer made from ZN catalysts (~160–162°C). This is due to lower crystallinity. Heat sealing film applications can take advantage of lower melting points.
Molecular Weight Distribution	MWD is narrow and clearly differentiates mPP from ZN PP ex reactor. Narrow MWD impacts many physical properties, For example, it is good for fibers processing with respect to filament continuity, fine denier fiber, and lower lint from melt blown fibers.
Organoleptics	Reduced content of extractable or volatile low MW material vs ZN random copolymer resins [33] reduces trace taste or odor characteristics, and also reduces smoke during resin processing.
Polymer residues	SSC supports are usually metal oxides, not metal halides as with ZN catalysts. Thus, there are less acidic residues and less neutralizer additive is needed for the polymer.
Purity	Peroxides are not needed for visbreaking in order to narrow MWD, increasing resin purity. This is important for food or medical applications.
Surface Gloss	Resin uniformity leads to smoother surfaces and excellent gloss.

(Continued)

Table 6.2 (cont.) Some characteristics of mPP resins.

Resin Property	Characteristics and Comments
Tacticity Range	The complete range of microstructure tacticities are achievable. However, the most notable is syndiotactic PP, which is not accessible from commercial ZN catalysts.
Transparency	Crystalline domains are smaller than ZN PP, which improves clarity. Also, nucleating agents used to reduce crystallite size are more effective due to mPP resin homogeneity [53]. This reduces haze and gives high transparency.

6.7   Selected Applications of mPP Resins

To give the reader a sense of the breadth of possibilities for mPP, several applications are described below. The intent is to highlight how mPP properties are used commercially.

6.7.1   Medical Applications

In medical applications, such as syringes or titration plates, stiffness, clarity, barrier properties and purity are important desirable parameters. Light weight and low cost (*i.e.* disposability) are also important. Due to these requirements, conventional PP grades are in wide use in medical applications. However, mPP provides some interesting advantages. Due to its more narrow MWD, mPP has an increased rate of crystallization, which leads to smaller crystallite size. Consequently, light interacts less with mPP and provides better clarity, and the clarity can be maintained over a stiffness range where Z-N resins are quite hazy. This also reduces the use of nucleation or clarification additives. The elimination of low MW and high MW fractions reduces distortions during molding and warping of molded parts, and reduces extractable components. The higher purity of mPP imparts better thermal stability during sterilization or radiation, reducing yellowing. Thus, mPP is an interesting, globally available material for medical usage [54, 55, 56].

6.7.2   Thin Wall Food Containers

Food contact applications require that the container imparts negligible odor and taste into the contained food. Good container transparency enhances consumer response leading to increased sales. Container impact resistance improves packaging durability on the shelf. New mPP grades may compete effectively with ZN resins, which require post reactor blending with plastomers and are subject to inhomogeneities. The absence of vis-breaking additives in mPP improves organoleptic characteristics [57].

6.7.3   High Clarity Bottles for Personal Care

Random copolymers of mPP, with low warpage and excellent clarity and gloss lend themselves toward production of bottles for personal care products such as cosmetics where the package is integral to the product. In such applications these resins compete with polyethylene terephthalate (PET) which is known for its glass-like clarity and gloss. Nucleating agents may be added to increase clarity. The lower density of mPP ($0.9 \, \mathrm{g/cm^3}$) vs. PET ($1.4 \, \mathrm{g/cm^3}$) provides a weight saving, and mPP can be processed at lower temperature than PET [58].

6.7.4   Waxes

Low MW, isotactic polypropylene waxes with narrow MWD, tunable melting points and hardness can be prepared from certain bridged metallocenes and low polymerization temperatures. They have a variety of specialty uses. The catalyst and process conditions promote the formation of terminal unsaturation useful for subsequent functionalization. For example, direct reaction of the wax with maleic anhydride gives a resin that is used to improve paint adhesion in TPOs [59]. Copier toners are prepared using low MW iPP in combination with styrenic-acrylic resins, carbon black (or colorant) and other ingredients. These toners exhibit improved hardness and transfer efficiency to the paper. Low MW mPP waxes are useful as dispersing aids for colorants in high MW PP and as matting agents in powder coatings [60, 61].

6.7.5 New Resins from Hybrid Catalysts to Make Polypropylene Alloys

It is possible to prepare new resins by combining a ZN supported Ti catalyst and metallocene catalyst into one hybrid system. Subsequent polymerization produces "alloys", *i.e.* intimate mixtures of polymer types produced by different catalyst types [62]. These are superior to mechanical blends which are less uniform on a microscopic scale. In one report the hybrid is prepared in two stages: the supported ZN catalyst is prepared by traditional means (Chapter 4), and in the second stage that catalyst is slurried with a solution of a metallocene and triethylaluminum to reduce the metallocene and drive it to the surface of the support [63]. The catalyst is then employed in a multistage process. In the first stage homopolymer is prepared using TEAL/organosilane cocatalyst. Then MAO, as a solution in toluene is introduced into the reactor, activating the metallocene, and an elastomeric phase is produced using a mixture of propylene and ethylene. The distribution of the ethylene monomers in the polymer is more random (less blocky) than that of a ZN catalyst, giving it superior elastomeric properties.

6.8 Metallocene Synthesis

The beauty and complexity of the synthesis of metallocenes are attractive to synthetic chemists. However, practical scale-up and manufacture are quite challenging. Multiple steps, specialized reagents, cryogenic temperatures, and recrystallizations may be required. Syntheses more resemble pharmaceuticals manufacture than industrial Ziegler Natta catalyst production [64]. One recent example from a family of metallocenes useful for high molecular weight polypropylene and copolymers demonstrates this point [51]. Figure 6.6 outlines the multistep synthesis of this metallocene. The described route to the final supported catalyst requires about 75 synthetic, washing, and separation steps, utilizing over 30 different chemicals shown in Table 6.3. Several of these reagents have pyrophoric properties, and many are quite expensive to purchase commercially. The amount of byproducts that have to be disposed of or recycled is also daunting compared

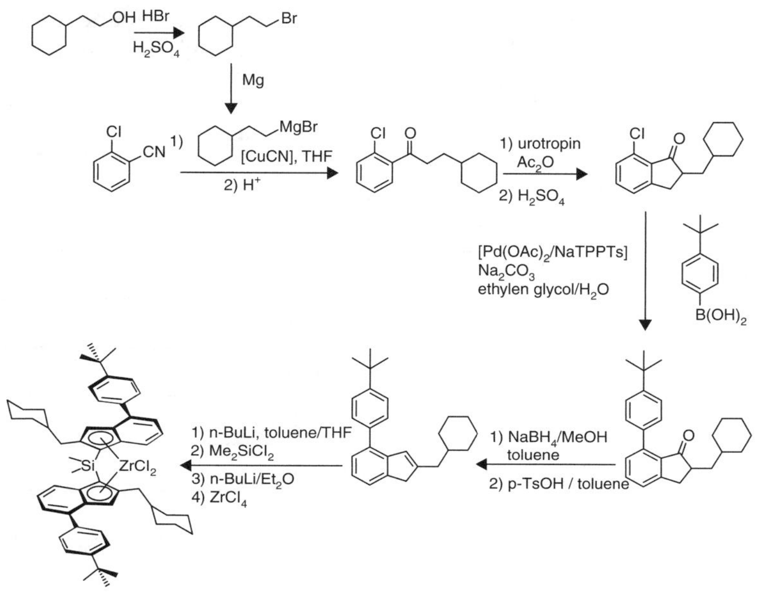

Figure 6.6 Synthesis of Dimethylsilandiyl-*bis*-(2-cyclohexylmethyl)-4-(4'-tertbutylphenyl)-1-indenyl-zirconium dichloride, Supporting Steps Omitted. Reproduced from US patent application 2010/0267907.

to ZN catalyst manufacture due both to its volume and complexity; the synthesis below generates over 30 separate streams of spent materials. While this example may not represent a metallocene currently in commercial use, it is indicative of the synthetic sophistication in the patent literature and the hurdles for commercialization.

Because a chiral metallocene is required for iPP, there is an additional synthetic complexity. The metallocene preparation can produce a mixture of racemic and meso isomers which must be separated by selective crystallization. Schematically this is shown in Figure 6.7, and, of course, it reduces the overall yield of the desired racemic isomers. In favorable cases, the meso isomer can be isomerized to a further mixture of racemic and residual meso metallocenes [65]. There have also been developments to recover ligands from the synthetic by-products using preparative HPLC [66].

Table 6.3 Reagents used for the preparation of dimethylsilandiyl-bis-(2-cyclohexylmethyl)-4-(4′-tertbutylphenyl)-1-indenyl-zirconium dichloride catalyst on silica [41].

2-Cl-benzonitrile	Mg turnings*
2-cyclohexyl ethanol	$MgSO_4$
4-t-butyl benzene boronic acid	Na_2CO_3
Acetic anhydride	$NaBH_4$
CH_2Cl_2	NaCl
CH_3OH	$NaHCO_3$
CuI	NaOH
Et_2O	n-BuLi*
H_2SO_4	NH_4Cl
HBr	$Pd(OAc)_2$
HCl	Pentane
Heptane	p-toluene sulfonic acid
Hexamethylenetetraamine	SiO_2
I_2	THF
Isohexane	Toluene
Methylaluminoxane*	$ZrCl_4$

Pyrophoric properties

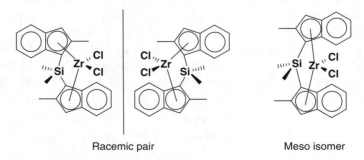

Racemic pair   Meso isomer

Figure 6.7 Racemic and Meso Metallocene Isomers.

6.9   Syndiotactic Polypropylene

Practical preparation of syndiotactic polypropylene (sPP) is a unique accomplishment of propylene polymerization by single site catalysis, and, accordingly, is discussed here as a separate topic. The structure of sPP is shown in Figure 2.2 and is conveniently described as a PP form with alternating placement of the methyl groups along the backbone. It has been known since the original work of Natta, and the best ZN catalyst system is based upon a vanadium catalyst and DEAC operating at low temperature [67]. However, this proved impractical for commercialization, and sPP remained a novelty until J. Ewen developed a practical SSC for sPP in 1988 [68, 69, 70]. The pentad structure by NMR is described as rrrr (see Figure 2.2 and Section 2.2.2.) and is often reported in the 85–95% range, compared to 90–99% mmmm pentad purity of iPP.

The most cited catalyst structures for sPP have a fluorenyl ligand (Figure 6.3f) and a single symmetry plane, and would not be expected to produce iPP. The labile coordination sites are enantiotopic and provide the opportunity for alternating methyl placements (*i.e.* alternating orientation of the incoming monomer). This is believed to be controlled by the bulky, rigid ligand structure directing the growing polymer chain orientation, which then controls the propylene orientation during insertion [71]. In the absence of that control, which is dependent upon the influence of the ligand sterics, atactic PP would result. Many types of fluorenyl ligand metallocenes have been prepared and studied, and a comprehensive review has been published [72].

Syndiotactic polypropylene is less crystalline than iPP, and it has smaller crystalline domains. This contributes to very good optical clarity and a lower melting point (~128°C) than iPP (~155–165°C) [73]. sPP films have a uniform smooth surface, high gloss and a soft feel. The resins have found use in a variety of specialized areas, such as heat seal film packaging applications where lower melting point and low temperature flexibility are distinct advantages. sPP has been found promising for power cable applications compared to several other materials [74]. It can be used as a minor component in iPP blends to improve impact strength and clarity at the expense of stiffness [75]. Alternatively, when it is desirable to have a resin with properties of both sPP and iPP two different metallocenes can be incorporated into the catalyst system to product a mixed resin *in situ* [76]. sPP is more resistant to UV radiation than iPP, and so it is interesting for medical applications such as sutures [77].

One drawback of sPP is that it crystallizes slowly. This has important practical effects, reducing production rates at fabricators and requiring added energy for cooling. The effect may be due to inter-conversion between two crystalline phases, of which the higher melting has a rubbery nature. Recent developments show crystallization speed can be improved by the use of certain additives [78, 79].

It is feasible to impart low syndiotacticity into iPP by refinements in metallocene design. By specific substitutions on the aromatic ligands and the *ansa* bridge a continuum of stereo defects can be introduced. Isotactic structures with 3–4% rr defects yield stiff polymers, whereas 4–6% rr gives more flexible thermoplastics, and 7–11% result in thermoplastic elastomers [80, 81]. As the % rr increases the melting point decreases, reflecting smaller crystalline domains, and a shift from the alpha to the delta crystalline form. The latter tolerates rr defects more readily.

6.10 Commercial Reality and Concluding Remarks

As of the publication of this text, SSCs for mPP have achieved only modest commercial success [82, 83]. About 1.4 million metric tons of mPP was estimated to have been sold in 2009, accounting for <3% of global PP sales [84]. This is a sizeable increase from the <0.5% of market in 2002 but still represents only a small penetration in the last decade [85]. ExxonMobil, Fina and LyondellBasell are among the leading mPP producers.

There are a number of reasons for the limited market penetration of this advanced science. First and foremost, ZN catalysts are well-entrenched, successful and very economical. A forte of most commercial ZN catalysts is that a limited catalyst product range is capable of producing a wide range of high quality resins. This is a relatively weak point of metallocene catalysts which are more individually tailored for specific resin types [71]. ZN catalyst capabilities and versatility have been improved through continued developments in external donor technology (see Chapter 4). For metallocenes, up to the present, external donors have not proven useful.

A second important factor is that the cost of single site catalyst systems is higher than Ziegler Natta catalyst systems [86]. The complexity and expense of metallocene synthesis is described in section 6.8. In addition, MAO cocatalyst is more costly by about an order of magnitude than conventional alkylaluminum cocatalysts, such as TEAL, and often MAO is used in large excess compared to the transition metal component. To the extent that activated supports (Section 6.4.3) are commercially viable, this factor is mitigated. However, it is believed that activated supports are not in significant commercial use as of this writing.

A third factor is that the commercial application of metallocenes and MAO relies upon the infrastructure of existing polypropylene plants. This creates logistical issues for handling of the catalysts and cocatalysts separate from ZN catalysts, cross-contamination issues, and the transition between resin grades in continuous reactors [87].

Despite these obstacles, single site catalysis plays a useful and growing role in the polypropylene industry. The utility will not be in the workhorse homopolymer, random and impact copolymer resins now successfully served by Ziegler-Natta catalysts and for which there is little incentive for change. Rather it will be in new specialty resins, where resin properties are achievable only with a SSC. Exciting developments continue to unfold, and in the coming years one can expect additional markets will develop for mPP.

6.11   Questions

1. Why would metallocene systems for producing atactic polypropylene function well in the absence of a catalyst support?
2. A silica support is slurried in a toluene solution of MAO and then isolated by filtration and dried without toluene washing. What might be the consequence of omitting the washing step?
3. In hybrid catalysts containing both ZN and SSC sites, what effect might the internal and external donors have on the SSC sites?
4. One of the literature methods for preparation of a supported metallocene is prereaction of a metal alkyl with a hydrated support, followed by reaction with the metallocene [88]. What practical advantage would this provide?

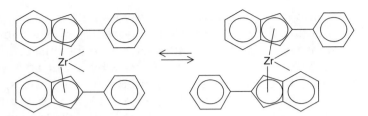

Figure 6.8 Metallocene with interconverting structures.

5. A certain method for supporting an *ansa* metallocene, (Cp–Cp)ZrMe$_2$, is by physically depositing the metallocene by precipitation from solution onto the support with a very low hydroxyl content. What might the result following activation with a cocatalyst such as MAO?

6. The metallocene in Figure 6.8 can interconvert between the two isomers shown [89]. What types of polymer might be produced from each isomer?

7. What steric factor is unique about the structure in Figure 6.3c) compared to the other structures in Figure 6.3?

8. One novel method of producing MAO is to use hydrated metal salts in place of water. This moderates the exothermic reaction and may provide improved control. From a practical perspective, suggest several process development issues that may impede commercialization.

References

1. D.S.Breslow, N.R. Newburg, *J. Am. Chem. Soc.* vol. 79, pp. 5072–73, 1957.
2. K. H. Reichert. K. R. Meyer. *Makromol. Chemie*, vol. 169, p. 163, 1973.
3. H. Sinn, W. Kaminsky, *Adv. Organomet. Chem.* vol. 18, 99–149, 1980.
4. Stereoselective Polymerization with Single Site Catalysts, L.S. Baugh and J.A.M. Canich editors, CRC Press, Boca Raton, FL, 2008.; *Polypropylene Handbook*, 2nd edition, Nello Pasquini editor, Hanser Publishers 2005.; R. Leino, *Encylopedia of Polymer Science and Technology*, John Wiley and Sons, Inc., pp. 136–179, 2001.; H.H. Brintzinger, D. Fischer, R. Miilhaupt, B. Rieger, R.M. Waymouth, *Angew Chemie*, Int. Ed. Engl., vol. 34, pp. 1143–1170, 1995.
5. For general reviews of metallocene chemistry of all metals see a) *Metallocenes*, J. Long, Blackwell Science, Inc,. Malden MA, 1998 b) *Metallocenes*, A. Togni, R.L. Halterman, Eds, Wiley-VCH New York, 1998.c) ACS Symposium #857: *Beyond Metallocenes: Next-Generation*

Polymerization Catalysts Edited by G.G. Hlatky, A.O. Patil, A.C. S, Washington, D.C. 2003. d) D.H. Camacho, Z. *Guan.Chem. Commun.*, 2010, 46, 7879–7893.

6. The term "hapto" describes a multidentate liqand of contiguous atoms coordinated to a central atom. The superscript indicates the number of contiguous atoms bound to the central atom, which is usually a metal.

7. W. Spaleck, F. Kuber, A. Winter, J. Rohrmann, M. Antberg, V. Dolle, E.F. Paulus, Organometallics, vol. 13, p. 954, 1994; W. *Kaminsky Angew. Makromol. Chem.* vol. 223, p. 101, 1994; H.H. Brintzinger, D. Fischer, R. Mulhaupt, B. Rieger, R.M. Wamouth, Angew. Chem. Int. Ed. Engl., vol. 34, 1143; J.A. Ewen, R.L. Jones, A. Razavi, J.D. Ferrara, *J. Amer. Chem. Soc.*, vol. 110, 1988; R. Lieberman, "Propylene Polymers", *Encylopedia of Polymer Science and Technology*, vol. 11, pp. 287–358; C. De Rosa, F. Auriemma, *Polymer Chemistry* (RSC), DOI: 10.1039/c1py00129a, June, 2011.

8. For background see: Molecular Symmetry by D.J. Willock, J. Wiley and Sons, 2009; *Stereoselective Polymerization with Single Site Catalysts*, L.S. Baugh and J.A.M. Canich editors, CRC Press, Boca Raton, FL.

9. C2 denotes a 180° rotation axis, whereas C1 has only the identity, 360° rotation axis.

10. K. Kawai, M. Yamashita, Y. Tohi, N. Kawahara, K. Michiue, H. Kaneyoshi, R. Mori, US Patent No. 7449533.

11. V. Marin, A. Razavi, US Patent No. 7470759.

12. J. Schottek, N.S. Paczkowski, A. Winter, T. Sell, US Patent No. 7285608.

13. A.Z. Voskoboynikov, A.N. Ryabov, M.V. Nikulin, A.V. Lygin, D.V. Uborsky, C.L. Coker, J.A.M. Canich, US Patent No. 7868197.

14. T. Yamaguchi, J. Tanaka, S. Otsuzuki, Y. Tohi, K.Nagahashi, N. Yamahira, S. Ikenaga, S.K. Moorthi, K. Kamio, US Patent App. No. 2010/0069588.

15. L.R. Sita, W. Zhang, US Patent App. No. 2011/0028654.

16. F. Kuber, B. Bachmann, W. Spaleck, A. Winter, J. Rohrmann, US Patent No. 6051727.

17. R. Leino, S. Lin, *Isotactic Polypropylene from C2 and Pseudo C2 Symmetric Catalysts*, Stereoselective Polymerization with Single Site Catalysts, L.S. Baugh and J.A.M. Canich editors, CRC Press, Boca Raton, FL, p 19, 2008.

18. For recent examples see. a) Ernst, K. Hakala. P.Lehmus, US Patent No. 7915367; b) H. Gregorius, V. Fraajie, M. Lutringhauser, US Patent Application 2009/0030166A1.

19. K. Press, A. Cohen, I. Goldberg, V. Venditto, M. Mazzeo, M. Kol, Ang. Chemie, Int. Ed., vol. 50, pp. 3529–3532, 2011; A. Razavi, V.P. Marin, M. Lopez, US Patent No. 7649064.

20. V. C. Gibson, S. K.Spitzmesser, *Chem. Rev.*, vol. 103, pp. 283–315, 2003.

21. D.L. Deavenport, J.T. Hodges III, D.B. Malpass, N.H. Tran, US Patent No. 5206401; J.K. Roberg, R.E. Farritor, E.A. Burt, US Patent No. 5606087.

22. E. Zurek, T. Ziegler, *Progress in Polymer Science*, vol. 29, pp. 107–148, 2004.
23. A.R. Barron, *Organometallics*, pp. 3581–83, 1995.
24. R. Leino, *Encyclopedia of Polymer Science and Technology*, John Wiley and Sons, Inc., pp. 136–179, 2001; R. Lieberman, "Propylene Polymers", *Encyclopedia of Polymer Science and Technology*, vol. 11, pp. 287–358; A.E. Hamielec, J.B.P. Soares, *Polypropylene: An A-Z Reference*, J. Karger-Kocsis editor, Kluwer Publishers, Dordecht, pp. 447–453, 1999.
25. There is a potential that two olefins may coordinate simultaneously, as pointed out by one reviewer. See Q.Yang, M. D. Jensen, M. P. McDaniel, *Macromolecules*, vol. 43, pp. 8836–8852, 2010.
26. D. Malpass, Introduction to Industrial Polyethylene: Properties, Catalysts, and Processes,. Scrivener Publishing LLC, John Wiley and Sons, Inc, 2010.
27. H. Gregorius, V. Fraajie, M. Lutringhauser, US Patent Application 2009/0030166.
28. P.D. Jones, D.B. Malpass, E.I. Band, G.M. Smith, B.L.S. Hudock, US Patent No. 6046347; G.M. Smith, D.B. Malpass, S.W. Palmaka, US Patent No. 5731451; G.M. Smith, D.B. Malpass, US Patent No. 5728855; S.A. Sangokoya, US Patent No. 5731253.
29. L. Lubin, S.A. Sangokoya, J.R. Strickler, S.P. Diefenbach, US Patent No. 7960488.
30. H.W. Turner, G.G. Hlatky, R.R. Eckman, US Patent No. 5384299.
31. J.A.M. Canich, H.W. Turner, G. Hlatky, US Patent No. 7163907.
32. M.P. McDaniel, J.D. Jensen, K. Jayaratne, K.S. Collins, E.A. Benham, N.D. McDaniel, P.K. Das, J.L. Martin, Q. Yang, M.G. Thorn, A.P. Masino, Metallocene Activation by Solid Acids, in *Tailor-Made Polymers*, Ed. J.R. Severn and J.C. Chadwick, Wiley-VCH Verlag GmbH & Co., Weinheim, Germany, 2008, pp. 171–210, 2008.
33. G.R. Hawley, M.P. McDaniel, M.D. Jensen, C.E. Whittner, US Patent No. 6573344; G.R. Hawley, M.P. McDaniel, C.E. Whittner, M.D. Jensen, J.L. Martin, E.A. Benham, A.P. Eaton, K.S. Collins, US Patent No. 7109277; M.G. Thorn, Q. Yang, K.C. Jayaratne, US Patent No. 7501372; A. Razavi, V.P. Marin, M. Lopez, US Patent Application No. 2009/0156761 A1.
34. L.A. Williams, T.J. Marks, *ACS Catalysis*, vol. 1, 238–245, 2011.
35. S.L. Wegener, T.J. Marks, P.C. Stair, *Accts. Chem. Res.*, published on line, DOI 10.1021/ar2001342, 2011.
36. Y. Nakayama, J. Saito, H. Bando, T. Fujita, *Chem. Eur. J.*, vol. 12, PP. 7546–7556, 2006.
37. Use of zeolites: Q.C. Bastos, M.D.F. V. Marques, *J. Polym. Sci. A Polym. Chem.* vol 43, pp. 263–272, 2005.
38. Use of clay: H. Ohtaki, N. Iwama, M. Kashimoto, T. Kato, T. Ushioda, US Patent No. 7906599.
39. Use of clay: K.Y. Shih, D.A. Denton, M.J. Carney, US Patent No. 6559090.

40. Use of polystyrene: T. Kitagawa, T. Uozumi, K. Soga, T. Takata, *Polymer*, vol. 38, pp. 615–620, 1997.
41. J.CW. Chien, B.P. Wang, *J. Polym. Sci. Part A, Polym. Chem.*, vol. 26, p. 3089, 1988; J.CW. Chien, B.P. Wang, *J. Polym. Sci. Part A, Polym. Chem.*, vol. 28, p. 15, 1990.
42. A.E. Hamielec, J.B.P. Soares, *Polypropylene: An A-Z Reference*, J. Karger-Kocsis editor, Kluwer Publishers, Dordecht, pp. 447–453, 1999.
43. G. Fink, B. Steinmetz, J. Zechlin, C. Przybyla, B. Tesche, *Chem. Rev.* vol. 100, pp. 1377–1390, 2000; Y. Choi, J.B.P. Soares, *Can. J. Chem. Eng.*, vol. 9999, published on line, DOI 10.1002/cjce.20583, 2011.; S.L. Wegener, T.J. Marks, P.C. Stair, Accts. *Chem. Res.*, published on line, DOI 10.1021/ar2001342, 2011.
44. See Y. Choi, J.B.P Soares, ref. 37 for classification of techniques including a description of spacer groups.
45. A. Hanyu, R. Wheat, "Properties and Applications of Metallocene – Based Isotactic Polypropylenes", *Metallocene Technology in Commercial Applications*, G.M. Benedict editor, William Andrew Publishing, Plastics Design Library, pp. 101–109 (1999).
46. J.A. Ewen, US Patent No. 7332456.
47. C. De Rosa, F. Auriemma, R.C.S., *Polymer Chemistry*, 2011, http://pubs.rsc.org | doi: 10.1039/c1py00129a
48. Young's modulus is the slope of the stress:strain curve and is a measure of polymer's stiffness.
49. T.C. Yu, D.D. Metzier, *Handbook of Polypropylene and Polypropylene Composites*, 2nd edition, H.G. Karian editor, Marcel Dekker, New York, pp. 201–250, 2003.
50. P.K. Hanna, D.D. Truong, US Patent No. 7183359.
51. A. Dimeska, R.D. Maier, N.S. Paczkowski, M.G. Thorn, A. Winter, J. Schulte,T. Sell, US Patent Application No. 2010/0267907.
52. See for example, http://www.ptonline.com/articles/the-new-polypropylenes-they've-got-more-of-everything, "The New Polypropylenes - They've Got More of Everything", L. M. Sherman, *Plastics Technology*, May 2002.
53. J. Rosch, J.R. Grasmeder, "Transparaent Metallocene-Polypropylene for Injection Molding", *Metallocene Technology in Commercial Applications*, G.M. Benedict editor, William Andrew Publishing, Plastics Design Library, pp. 147–152 (1999).
54. J. Welch, *Business Briefing: Medical Device Manufacturing and Technology*, pp. 63–66, 2004.
55. B. Schutz, R. Konrad, *Business Briefing: Medical Device Manufacturing and Technology*, pp. 1–4, 39–41, 2005.
56. *Purell* HM671T Product Information Sheet, Basell Polyolefins, 2006. *Petnology magazine*, Aug. 14, 2007; http://www.petnology.com/zine/zine.php?c_ID=10962&ilang=e

57. *Petnology, magazine* Jan. 13, 2010; http://www.petnology.com/zine/zine.php?c_ID=11819&ilang=e&skw=metallocene&lang=e
58. *Petnology, magazine* Jan. 13, 2010; http://www.petnology.com/zine/zine.php?c_ID=11819&ilang=e&skw=metallocene&lang=e
59. P. Hanna, S. Nistala, L.L. Otte, K. Fudge, T.J. Clark, D.D. Truong, J. Woods, US Patent No. 7589150.
60. R. Hess, *Plastics, Additives and Compounding*, vol. 4 (5), pp. 28–31, 2002.
61. Clariant Ceridust 6050M. http://www.clariant.com/corp/internet.nsf/directname/ecs
62. P. Galli, *Macromolecular Symposia*, vol. 112, Issue 1, p. 1–16, 1996.
63. L. Lu, H. Fan, B.G. Li, S. Zhu, *Ind. Eng. Chem. Res*. vol. 48, 8349–8355, 2009.
64. In fall 2011 Grace Davison purchased Synthetech, a producer of chiral and peptide pharmaceutical intermediates.
65. R.W. Lin, US Patent No. 5965759.
66. P. Muller, R.L. Jones, R. Chevalier, C. Sidot, V. Garcia, US Patent No. 7951970.
67. A. Zambelli, E. Bajo, E. Rigamonti, *Makromol. Chem.* vol. 179, p 1249, (1978).
68. J. Ewen, A. Razavi, US Patent No US4892851.
69. J.A. Ewen, J.A.C.S., vol 106, p. 6355, 1984.
70. J.A. Ewen, R.J. Jones, A.Razavi, J.D. Ferrara, *J.A.C.S.*, vol. 110, p. 6255, 1988.
71. E.P. Wasserman, "Metallocenes", *Encyclopedia of Polym. Sci. and Tech.*, J. Wiley and Sons, vol. 7, p. 73, and references therein.
72. C.J. Price, L.J. Irwin, D.A. Aubry, S.A. Miller, *Fluorenyl Containing Catalysts for Stereoselective Polymerization*, pp. 37–82, Stereoselective Polymerization with Single Site Catalysts, L.S. Baugh and J.A.M. Canich editors, CRC Press, Boca Raton, FL, 2008.;
73. However, one recent publication describes a metallocene catalyst that produces sPP with a m.p. = 171°C. See V. Marin, A. Razavi, US Patent No.7538167.
74. K.Yoshino, A. Ueda, T. Demura, Y. Miyashita, Y. Kurahashi, Y. Matsuda, Proceedings of the 7th International Conference on Properties and Applications of Dielectric Materials, vol. 1, pp. 175–178, 2003.
75. J. Schardl, L. Sun, S. Kimura, R. Sugimoto, *J. Plastic Film and Sheeting*, vol. 12, p. 157, 1996.
76. A. Razavi, US Patent App. 2006/0241254.
77. C.K. Liu, US Patent No. 5269807.
78. D.L. Dotson, US Patent No. 6642290.
79. V. Barre, L. Kelly, L. Lumus, M.B. Miller, J.M. Schardl, Jr. US Patent No. 6844381.

80. C. De Rosa, F. Auriemma, A. Di Capua, L. Resconi, S. Guidotti, I. Camurati, I. E. Nifant'ev, I.P. Laishevtsev, *J. Am. Chem. Soc.* vol. 126, 17040–17049 (2004).

81. C. De Rosa, F. Auriemma, *J. Am. Chem. Soc.*, vol. 128, 34, 11024–11025 (2006).

82. E.P. Wasserman, *Encyclopedia of Polym. Sci. and Tech.*, J. Wiley and Sons, vol. 7, p. 93.

83. L. Resconi, C. Fritz, *Polypropylene Handbook*, 2nd edition, Nello Pasquini editor, Hanser Publishers 2005, p. 107.

84. A.H. Tullo, C&E News, "Metallocenes Rise Again", vol. 88, (42), pp. 10–16, Oct. 18, 2010.

85. http://www.sriconsulting.com/PEP/Public/Reports/Phase_99/RP128C/, Process Economics Program Report 128C, Polypropylene Update 128C, November 2002.

86. P. Arjunan, Advantage™ Ziegler Natta PE Catalyst for Solution LLDPE Process Platform", ACS Poly Div Workshop Conference: "Advances in PO 2011", Sep 25–28, 2011, Sonoma Valley, CA, USA.

87. R. Tharrappel, R. Oreins, W. Gauthier, D. Attoe, K. McGovern, M. Messiaen, D. Rauscher, K. Hortmann, M. Daumerie, US Patent No. 7960484.

88. M. Chang, US Patent No. 5006500.

89. G.E.W. Coates, R.M. Waymouth, *Science*, vol. 267, p. 217, 1995.

7

Catalyst Manufacture

7.1 Introduction

From the preceding chapters it should be clear to the reader that the propylene polymerization catalyst is the centerpiece of propylene polymerization technology. It strongly influences the polymerization process design and defines the attainable polymer properties. So, the manufacture of high quality, consistent catalyst is critical to the successful operation of the polypropylene plant. Unlike the polypropylene plant, which is always a continuous process, catalyst manufacture is usually a batch manufacturing process. This chapter provides a concise overview of that process.

7.2 Development of the Manufacturing Process

A very large amount of research goes into developing a successful catalyst recipe. Almost all of this research is conducted on a laboratory scale, where less than a hundred grams of catalyst is prepared in a single synthesis. The laboratory development program of the catalyst synthesis optimizes results primarily in terms of polymerization performance, not in terms of efficient catalyst synthesis. Once performance goals are achieved the catalyst recipe is optimized, but within the constraints of the benchmark performance.

Confirmation of the optimized catalyst recipe requires production of larger amounts of catalyst, measured in kilograms in a pilot operation. The purpose of the pilot operation is confirmation of the polymerization performance and polymer quality, and the catalyst recipe is kept essentially unchanged for this. Hundreds of thousands, or possibly millions, of pounds of polypropylene will be produced before the catalyst recipe is considered successful and ready for commercial manufacture. Once optimized and confirmed, changing the catalyst recipe for convenience of manufacture is not a priority. Small changes may be made due to the necessities of large scale production, but the priority is to maintain the catalyst quality that has been demonstrated, and that is achieved by maintaining the integrity of the catalyst recipe, as far as practical. So, the commercial catalyst plant is ultimately designed around the successful, reproducible laboratory recipe [1]. This is simply because the value of the polymer far exceeds the savings that might be obtained by compromising the catalyst recipe to save manufacturing cost. Thus, catalyst manufacture is not necessarily an efficient process.

It is for the same reason, *i.e.*, the value of the polymer far exceeds the value of the catalyst, that almost all new catalyst development is done by the polypropylene producers themselves [2]. These same companies concurrently develop polymerization processes tailored to their catalysts. Thus, most modern propylene polymerization catalyst manufacture is conducted by polypropylene producers with proprietary processes. Often these producers license their catalysts and polymerization processes to other companies, so the catalyst plant may serve both the licensor and the licensee.

7.3 Chemistry of Catalyst Manufacture

The function of the catalyst plant is specialty chemical synthetic chemistry. Quality and consistent catalyst performance are crucial objectives. The principal reactions involved are 1) support formation, 2) support activation, and 3) catalyst isolation. Some examples of support formation include:

$$Mg(OCH_2CH_3)Cl + TiCl_4 \rightarrow MgCl_2 + (CH_3CH_2O)TiCl_3 \qquad (7.1)$$

$$MgCl_2 + nCH_3CH_2OH \rightarrow MgCl_2 \cdot (CH_3CH_2OH)_n. \qquad (7.2)$$

$$MgR_2 + xsSiCl_4 \rightarrow MgCl_2 + 2RSiCl_3 \qquad R = alkyl, xs = excess \qquad (7.3)$$

$$MgR_2 + \quad \diagdown\!\!\!\!\!\underset{\diagup}{Si}-OH \quad \longrightarrow \quad \diagdown\!\!\!\!\!\underset{\diagup}{Si}-O-MgR + RH \qquad (7.4)$$

The chemistry of support activation is a combination of reactions. It includes the extraction of portions of the support to create a more porous $MgCl_2$ matrix [3],

$$MgCl_2 \cdot (CH_3CH_2OH)_n + xs\ TiCl_4 \rightarrow$$
$$n(CH_3CH_2O)TiCl_3 + MgCl_2 \qquad (7.5)$$

the deposition of the internal donor onto surface sites on the $MgCl_2$ crystallites,

$$MgCl_2 + donor \rightarrow MgCl_2 \bullet donor$$
$$donor = phthalate\ esters,\ succinates,\ ethers,\ etc. \qquad (7.6)$$

the selective extraction of surface bound donor by solvents,

$$MgCl_2 \bullet donor + xs\ TiCl_4 \rightarrow MgCl_2 + TiCl_4 \bullet donor \qquad (7.7)$$

and, finally, the selective deposition of $TiCl_4$ onto surface sites of the $MgCl_2$ crystallites:

$$MgCl_2 + TiCl_4 \rightarrow MgCl_2 \bullet TiCl_4 \qquad (7.8)$$

In many cases, $TiCl_4$ functions both as the solvent for the selective extraction of surface bound donor, and for its own deposition onto the support surface.

The catalyst is isolated by thorough washing in purified hydrocarbon to remove excess $TiCl_4$ and residual dissolved complexes.

It is then dried to a powder under carefully controlled conditions to preserve particle morphology and drummed under anaerobic conditions.

7.4   Raw Materials Storage and Handling

The modern catalyst manufacturing plant infrastructure begins with large storage tanks for solvents and $TiCl_4$. On a weight basis, these are the plant's principal raw materials. The $TiCl_4$ storage tank is constructed of carbon steel and should be raised above ground on pedestals or piers to allow full access to inspect the surfaces. It must be diked to contain any spill, and the dike construction needs to account for the high density of $TiCl_4$ (1.73 g/mL). If there is a sump system inside the dike, it must not be connected to a drain, and rain water cannot be allowed to collect inside the dike. A foam or oil vapor suppression system for spills is required (*vide infra*) [4].

There may be separate storage tanks for recovered solvents. This depends upon the quality of the recycled solvents and the tolerance of the catalyst process to impurities. There is smaller storage for the catalyst support or its precursor, which may be a solid such as magnesium chloride or magnesium ethoxide, or a liquid such as a hydrocarbon solution of a magnesium alkyl. Less commonly, spherical silica may be a raw material [5], used as a templating agent for the active magnesium chloride support, or even a microspherical fumed silica seeding agent [6]. There is also storage for the internal donor(s), which also may be solid or liquid. Solid support precursors and the internal donors are usually received in drums or totes. Magnesium alkyls, if employed, are received as hydrocarbon solutions in large cylinders. All these materials are stored in a plant warehouse or a covered concrete pad. Metal alkyls storage should be segregated separately due to metal alkyls' pyrophoric nature. Liquids are piped directly to the process equipment. Drums of solids are hoisted into the unit where they are emptied.

The entire catalyst plant operates with the rigorous exclusion of moisture and air. Moisture in the process generates hydrochloric acid from hydrolysis of titanium tetrachloride. Moisture also produces hydrates of magnesium chloride, and carboxylic acids from hydrolysis of organic ester internal donors. Accordingly, nitrogen is utilized to blanket all operations, necessitating pipeline nitrogen or a liquid (cryogenic) nitrogen storage tank on site. If organometallics are used in the catalyst recipe, such as the use of magnesium alkyls as precursor to the $MgCl_2$ support, then there is also a strong sensitivity to oxygen. By contrast, the final supported catalysts are not usually susceptible to oxidation, because titanium is typically in the +4 oxidation state. Nevertheless, protection is still needed from air due to its moisture content.

Effective plant maintenance and safety require taking into account the extremely high corrosivity of the chemicals on site. It necessitates preventative procedures to preserve system dryness which minimizes corrosion to interior surfaces. Equipment is thoroughly flushed and dried before opening and purged with inert gas after maintenance. This may include the use of special corrosion resistant alloys, seal-less pumps and compressors.

7.5   Catalyst Preparation

Catalyst synthesis is conducted in a large, agitated reactor or series of reactors, with means to charge both powders and liquids. The first stage is the preparation of the catalyst support with a controlled morphology. This may require a specialized reactor and is the most demanding part of the catalyst synthesis. Its production requires tightly controlled reagent addition rates, special temperature profiles which may include refrigeration, precisely regulated agitation rate, and prolonged reaction time. Adjuvants, such as surfactants may be added to assist in achieving spherical particles [7]. Agitator design and the reactor's interior surfaces have a significant effect upon morphology; a poor design can cause particle attrition or agglomeration. Accordingly, seams and interior surfaces are polished. In some cases the support morphology is achieved by a spray cooling operation starting from molten magnesium compounds, such as alcoholates of magnesium chloride [8]. This requires sophisticated spray drying equipment and refrigeration [9]. Some examples of support formation operations are shown in Table 7.1.

The titanation of the catalyst support and the incorporation of the internal donor(s) are also demanding processes, but usually less so than the catalyst support-forming step. Often this part of the catalyst recipe will require several intermediate isolation steps of solid precursors to the final product. This necessitates some means of separation, which may be done by a large filter or by a decantation process. Filtration is, of course, a more efficient separation than

Table 7.1 Examples of mg- containing support precursor formation with controlled morphology for propylene polymerization catalysts.

Method	Example
Grignard [10]	Grignard is formed from Mg and slow addition of alkyl halide in the presence of tetraethoxysilane (TEOS) solvent. As the Grignard is formed, it alkylates the TEOS, forming a solid magnesium chloroethoxide support precursor.
Emulsion Cooling of $MgCl_2$ Solution [11]	A hot solution of $MgCl_2$ in ethanol is emulsified in Vaseline® oil and silicone oil by high speed agitation. The hot emulsion is slowly discharged into cold heptane, solidifying to a spherical support of $MgCl_2.3EtOH$.
Impregnation of Silica [12, 13]	Silica of narrow particle size distribution and high pore volume is mixed with a solution of dialkylmagnesium and then treated with a large excess of HCl gas.
Precipitation from a Nonalcoholic Solution of $MgCl_2$ [14]	$MgCl_2$ is dissolved in a mixed solvent system of toluene, tributyl phosphate and epichlorohydrin. The solution is cooled and solid support is prepared by slow addition of $TiCl_4$.
Solution Precipitation [10]	A solution of $MgCl_2$ and $Mg(OEt)_2$ is formed in an ether, and then precipitated under controlled agitation by addition of a hydrocarbon, to yield a solid magnesium chloroethoxide support precursor.
Magnesium Ethoxide [15]	Mg metal is suspended in a mixture of dry alcohols, predominantly composed of ethanol. A catalytic solution of iodine is added and $Mg(OEt)_2$ is formed.
Dealcoholation of a Magnesium Chloride Solution [16]	A $MgCl_2$ solution is formed in a mixed solvent of 2-ethylhexanol, alkane, dialkylether and ethyl silicate. Surfactant is added, the solution is cooled, and the support is formed by slow addition of $TiCl_4$.

decantation, and provides advantages in terms of more efficient washing operations. However, it requires a step to flush the solids from the filter back into the reaction train. It also adds to the plant investment and necessitates filter maintenance operations.

Milling and particle classification are two operations that are not commonly found in modern catalyst plants. By its nature, milling yields a product with minimal morphology control, and this has serious implications for the polymer producing plant. Classification, such as by sieving, was done to minimize that impact. Milling and classification operations were common with unsupported TiCl$_3$ catalysts and early supported catalysts, but those catalysts are largely obsolete today.

7.6 Catalyst Drying

Once synthesized, the final catalyst is isolated by a series of washing steps with an inert hydrocarbon, followed by drying in a nitrogen stream. This can be combined with reduced pressure to speed the drying process. It is crucial to design the drying stage so as to avoid breaking catalyst particles or clumping catalyst particles together. The former will result in polymer fines, and the latter will result in large clumps of polymer. Drying may be done by spraying the catalyst slurry into an inert atmosphere [17], or by passing a nitrogen stream through a bed of wet catalyst. In either case, a closed system may be used to capture the volatile solvent with a condenser and recycle nitrogen [18]. Nitrogen temperature is moderated to avoid cooking the catalyst particles; in practice the catalyst particle temperature is limited by the evaporative cooling of the solvent. Typically, about 5–15% solvent remains in the dry particles. The presence of the residual solvent in the catalyst pores provides some protection against contamination by adventitious moisture, but it also requires that the catalyst be handled as a flammable solid.

7.7 Catalyst Packaging

The final product is usually a free flowing, tinted powder. It is packaged in corrosion resistant, gas tight drums under a few pounds of nitrogen pressure. The drumming operation is done in a closed system that excludes contact with the atmosphere. Drum integrity is confirmed during storage by checking the pressure. Catalyst which is exposed to air may exhibit no visible change, even though performance has been impacted. This can happen very quickly because magnesium chloride is very hygroscopic. Concurrent with moisture exposure there will be a release of HCl fumes due to reaction with the supported $TiCl_4$. As exposure to moisture increases the catalyst may decolorize and take on a white appearance. The valving of the drum must allow for easy discharge of the catalyst, and it is mated to handling equipment in the catalyst charging section of the polypropylene plant.

In some cases catalyst batches are blended prior to packaging [19]. This smooths out modest performance differences that are inevitable from batch processes. Depending on the plant design, blending might instead be performed at the final catalyst slurry stage prior to drying. In some respects, a blending operation for batches of catalyst slurry is more convenient for the plant than is the processing required to blend dry catalyst powder. It is also possible to avoid drying the catalyst, and, instead, packaging it as a hydrocarbon slurry. In that case drum design and handling must take into account the need to resuspend the slurry when the container is discharged.

7.8 Recovery and Recycle of Spent Solvents

Even a casual reading of the polymerization catalyst literature reveals that catalyst manufacture consumes large amounts of titanium tetrachloride and hydrocarbon solvents relative to the amount of catalyst produced. One article has cited a spent liquids stream as over 30 times the weight of catalyst [20]. Another study estimated the variable cost associated with omitting solvent recovery at 158 Euros/kg catalyst, or approximately \$95/lb of catalyst [21]. So, for both economic and environmental reasons, the plant design must incorporate a large recycle operation to recover spent $TiCl_4$ and solvents. The facilities required for the solvent recovery can exceed the equipment required for the catalyst synthesis. Spent liquids must be stored separately from recovered solvents, and heated, agitated storage may be required to prevent precipation of suspended solids.

Solvent recovery and recycling is most commonly done by distillative separations [22]. Specifics of the design depend upon the nature and concentration of the byproducts, which, in turn, depend upon the specific catalyst recipe. For example, the separation of $TiCl_4$ (bp=136°C,) from *n*-hexane (bp=69°C) is simpler than $TiCl_4$ separation from heptanes (bp=98°C) or toluene (bp=111°C). Multi-plate distillation columns are used to effect the separations, often operated in a continuous mode. Operations at reduced pressure, by means of a vacuum system, can improve the overall distillate recovery.

It is highly likely that recovered solvents will introduce impurities into the catalyst synthesis, and these impurities can build up over time and repeated recycling. This requires careful study, and may result in recycling only a portion of spent solvents, or periodic venting operations for accumulated volatile impurities. For a multi-catalyst plant the situation is more complicated due to cross contamination possibilities.

The byproducts in the separations usually contain titanium alkoxyhalides, and titanium complexes with the internal donor or other Lewis bases, such as ethers. These may be present as suspended solids which can foul the distillation system with a viscous or solid residue. Therefore, it is necessary to retain a certain fraction of the solvents with the distillation bottoms to maintain flowability. Prolonged heating can increase the amount of tar. Much effort has been expended in the literature at devising efficient and practical separations of these types of mixtures [23].

The recovered solvents must be free of byproducts that can poison the catalyst or otherwise affect the stoichiometry of the process recipe. This includes the internal donor and derived decomposition products [24]. This is especially true of the solvent used for the final catalyst washings. If the catalyst plant makes more than one grade of catalyst, cross-contamination issues from minor components in the recovered solvents must be prevented, or separate recovered solvent storage tanks may be required.

After solvent recovery there is still a substantial bottoms fraction of byproducts. For example, in the conversion of magnesium ethoxide to magnesium chloride by reaction with $TiCl_4$, two moles of byproduct trichlorotitanium ethoxide are produced for every mole of magnesium chloride. Also, as a result of prolonged heating during fractional distillations, the byproducts can further react and decompose to become complex mixtures compounds with difficult handling properties. For the final distillation residues the main practical recourse is neutralization of acidity and subsequent disposal in an environmentally responsible manner. For many catalyst recipes the tonnage of this waste far exceeds the tonnage of catalyst product. Efforts to recycle spent materials into the process must be tempered by the necessity to have no impact on catalyst quality.

7.9   Prepolymerization at the Catalyst Manufacturing Plant

In some instances, the catalyst plant may also prepolymerize the catalyst [25]. Prepolymerization is discussed in Chapter 3. Catalysts that are prepolymerized are generally not dried. Rather, the final hydrocarbon slurry of the virgin catalyst is introduced into a pressure reactor where precise amounts of aluminum alkyl cocatalyst and external donor are added. Under gentle agitation about 5–20 weight equivalents of propylene are added while maintaining a low temperature until the polymerization is complete. The slurry is then transferred into shipping containers. The prepolymerized catalyst can be sensitive to storage time and temperature, and it can also be prone to settling and agglomeration. This is in contrast to the virgin catalyst, which is indefinitely storage stable, and resistant to clumping.

7.10   Plant Size

Catalyst plant capacity is in the range of a few hundred tons per year. This may seem very small, but it is comparable to other fine chemicals manufacture. For a modern, fourth or fifth generation supported catalyst, this capacity is sufficient to supply the needs of over ten world scale polypropylene plants. If modifications are made in the catalyst recipe that improve productivity, then, assuming catalyst plant capacity is unchanged, the ability to supply more polypropylene production capacity is increased. Thus, increasing catalyst activity reduces future capital investment.

7.11   Site Safety

The typical catalyst manufacturing facility stores large stocks of $TiCl_4$ and organic solvents. The former releases hydrochloric acid and titanium oxychloride dusts [26]. on exposure to moisture, whereas solvents, of course, are flammable. Fires can also arise from the catalyst powder, which contains flammable organics. In addition, suspended catalyst dusts are an explosion hazard. Accordingly, oxygen monitors are incorporated in the process to assure that flammable or explosive mixtures are avoided, and equipment is electrically grounded. This is especially true in the drying section of the plant where volatile organic solvents are stripped from the catalyst powder, and where the final catalyst powder is, itself, a flammable solid.

If the catalyst recipe includes the use of alkylmagnesium for support preparation or alkylaluminums for prepolymerization, then there is an additional fire hazard due to their pyrophoricity. Plant design should isolate, as much as is practical, the metal alkyl storage facility from other chemical storage. Special firefighting equipment is required on-site which uses dry chemical powders to smother fires. Water fog spray units may be installed to protect adjacent facilities. Special protective equipment is needed for personnel fighting fires, including aluminized suits and flame retardant coveralls. More information can be found in Chapter 5, and detailed discussions of metal alkyl fires can be found in the literature for the interested reader [27, 28].

For protection of the plant, its personnel and the surrounding community, the plant design includes sophisticated fire control equipment due to organic solvents, and $TiCl_4$ spill control systems. The latter includes automated alarms, diking of critical equipment and remote operated foam systems to blanket spills which would otherwise cause a voluminous, acidic smoke cloud. The acid cloud contains $HCl_{(g)}$ and an aerosol of $Ti(OH)_2Cl_2$ which can travel a considerable distance while slowly releasing additional HCl through the continued hydrolysis of the $Ti(OH)_2Cl_2$ [29]. The foam suppresses hydrochloric acid fumes, allowing emergency responders time to remediate the spill. Regular inspections are made for pools of rain water collecting around $TiCl_4$ tanks.

Chemical sampling is one of the more hazardous plant operations, because it exposes the operator to chemicals, and, furthermore, because these may be hot and/or under pressure. Thus, sampling protocols are a crucial aspect of plant safety. Careful consideration is given to protective gear, the sequence of opening and closing valves, purge samples, and taking final samples. At the same time, the purpose of sampling is to obtain material representative of the process, so the sampling point, sample size, and integrity of the sample container must be carefully devised. Liquid sampling may utilize septum-sealed, nitrogen-purged containers. The sample is taken from the process via a needle assembly directly into the sealed sample container to minimize exposure to the atmosphere and to the operator.

7.12 Quality Control and Specifications

Today's catalyst manufacturing plant includes a well equipped quality control laboratory. Incoming raw materials are screened for purity, especially water content, prior to use. High purity $TiCl_4$ is required because the presence of other transition metal contaminants might result in formation of unwanted catalytic centers [30]. Recycled materials are checked for specific contaminants. Each catalyst batch is manufactured under carefully controlled conditions and tested for both its composition and its polymerization performance in a laboratory test. Although many commercial fine chemicals are sold solely on the basis of compositional specifications, this is not the case with Ziegler-Natta catalysts. The catalyst is a complex mixture, containing a cocktail of active sites that may represent only a small fraction of the total material present. Thus, while compositional specifications are necessary, they are not sufficient to indicate performance. Composition parameters include the titanium, magnesium, halide, internal donor(s), and solvent contents. Catalyst physical properties measured include particle morphology and particle size distribution. The performance test measures the catalyst productivity under specified polymerization conditions (see Chapter 9), as well as basic properties of the polypropylene produced in the lab test (see Chapter 2). One of the interesting facets of catalyst quality control is the methodology required to get representative catalyst samples for the laboratory polymerization test. The catalyst batch size is hundreds of kilograms, but only a few milligrams are used for the lab polymerization. So, careful procedures are devised for the batch sampling, and in some cases duplicate testing may be done to assure quality. For example, sampling may be structured to obtain small quantities of catalyst during different phases of the packaging operation.

7.13 Diagram of a Hypothetical Plant

A schematic representation of a hypothetical catalyst plant is shown in Figure 7.1. Several storage tanks for virgin liquids are shown: titanium tetrachloride, toluene, and an aliphatic hydrocarbon. There is also drum or cylinder storage for other raw materials, including a Mg support precursor. Virgin and recycled solvent(s)

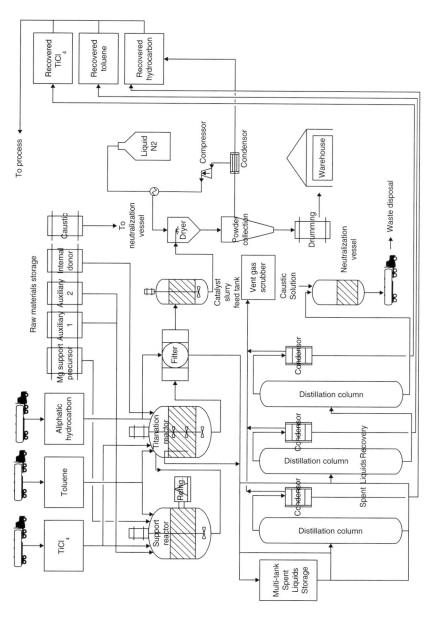

Figure 7.1 Hypothetical supported catalyst plant process diagram.

are added to the support preparation vessel. Then the remaining liquid and solid raw materials, including two auxiliary reagents needed to establish the correct support morphology, are metered into the support preparation vessel. Some of the reagents may be added incrementally during the support preparation. In this hypothetical plant, support preparation requires a refrigeration unit to moderate the solid support formation. The Mg support precursor is added by drums or totes. The support is prepared, and then the support slurry is transferred to the titanation reactor, where more solvents and the internal donor are added in a series of heating/settling/decanting steps. The supernatant liquids are discharged to the solvent recovery area. After the last titanation reaction the slurry is transferred to the filtration unit where the solids are collected and washed repeatedly with the aliphatic hydrocarbon. The spent hydrocarbon is directed to the solvent recovery system. The solid filter cake is resuspended in aliphatic hydrocarbon and transferred into a catalyst slurry feed tank, where it is slowly fed to the drying system. The dryer uses warm nitrogen, taken from a liquid nitrogen storage facility. The feed rates and residence time in the dryer determine the final residual solvent content in the catalyst. The powder catalyst is separated from the solvent-containing nitrogen gas in the powder collection vessel, and the former is passed through a condenser to recover the hydrocarbon solvent. From the collection vessel, the catalyst goes to a drumming facility and then to a warehouse.

The spent solvents, made up of various streams from the titanation, filtration, and washing steps are directed to a multicolumn solvent recovery system. The liquids can go directly to the columns or be fed to a storage tank. Recovered solvents are stored separately from the virgin solvents in this plant configuration. The final bottoms liquids collected from the columns are sent to a neutralizer where caustic is added, and the neutralized material shipped offsite for disposal.

7.14 Custom Manufacture

Catalyst manufacture is an intensive and expensive undertaking. There are instances where the polymer producer, having developed a catalyst, wants to maintain focus on its key business, *i.e.* polymer production. So, a manufacturing partner is sought to license the production of the newly developed catalyst. This will be a company that specializes in catalyst manufacture, and the endeavor is called custom catalyst or toll manufacturing. The custom catalyst manufacturer typically has extensive expertise and an existing infrastructure which can be utilized, at least in part, for the manufacture of a new, licensed catalyst. These arrangements are typically long term and require capital investment by the custom catalyst manufacturer. For example, in the 1990s the Amoco Chemical Company licensed the production of its supported catalyst to Catalyst Resources, a Houston, Texas custom catalyst company in a decade long agreement which justified the capital investment by Catalyst Resources [31]. Another manufacturing agreement was Himont and Mitsui's licensing of Stauffer Chemicals, which resulted in start up of a plant in Edison, N.J in the 1980s, and which continues operations in 2011. There are also custom catalyst manufacturers, such as Evonik, that offer a broad array of products of interest to the polymer producing industry. Evonik's portfolio of products, in addition to custom manufacture, includes external silane donors and catalyst support precursors [32]. Another catalyst manufacturer, Sud Chemie, offers propylene polymerization catalysts as a small fraction of a broad presence in catalyst manufacture, ranging from refinery catalysts to petrochemical catalysts to air purification catalysts [33].

Utilizing a custom catalyst manufacturer provides the licensing company several advantages. First, it allows the licensor to focus its energies and resources on polymer manufacture, which provides more revenue. Second, the catalyst manufacturer brings expertise and resources to the partnership. This can increase the speed of scale up and commercialization, and it also improves supply reliability on an ongoing basis. The custom manufacturer may also provide R&D support for catalyst improvements. Third, custom manufacture can provide the licensor an additional revenue stream. This is effected by allowing the custom catalyst manufacturer to sell the catalyst to third party polymer producers. These producers may be reluctant to deal directly with the catalyst licensor, who is likely to be a competitor. By working directly with the catalyst manufacturer under a confidentiality agreement, the third party polymer producer can disclose aspects of his process without that information flowing back to the original catalyst licensor. This provides synergy for licensing of the polymerization process. Recalling that a catalyst plant typically is sized to produce several hundred tons of catalyst, the catalyst licensor may not have the polymer production capacity to consume nearly that amount of catalyst. So, licensing to third parties helps to fill the plant capacity, reducing manufacturing costs for all involved.

7.15   Brief Consideration of Metallocene Catalyst Manufacture

The subject of metallocene catalysts is of minor, but growing importance for polypropylene, and is discussed in Chapter 6. The manufacture of metallocene catalysts generally involves two principal steps:

1. Preparation of the metallocene
2. Immobilization on a support

The first step is unlike Ziegler-Natta catalyst preparation which utilizes inexpensive $TiCl_4$ in large excess as the active catalyst center. Metallocene preparation more closely resembles the manufacture of pharmaceuticals. There are diverse steps peculiar to the individual metallocene with the use of expensive reagents. It is typically necessary to prepared several intermediates that are isolated by precipitation, extraction, and/or preparative chromatography. Yields can be moderate for individual steps, resulting in overall yields <50%.The exclusion of air is even more critical than for ZN catalysts because pyrophoric reagents are often used. The final metallocene is isolated by crystallization. The latter removes impurities and separates diastereomeric isomers. Most of the synthetic effort is devoted to construction of the ligand architecture. The scale of production is relatively small, being measured in the range of a few tons/year. Production quality control includes characterization by high field NMR or other sophisticated methods.

The second, supportation step does not involve preparation of a controlled psd magnesium support. Commercial silicas or other metal oxides are used, containing a specific hydroxyl content on the surface of the support. The supports used for manufacture of ZN polyethylene catalysts may be satisfactory. Binding the metallocene to the support can be done several ways which depend on how it is to be activated (see section 6.4). For example, in the case of an aluminoxane activator (section 6.4.1), the support may be reacted with an aluminoxane to bind it to the surface and then the metallocene added to bind it to the aluminoxane. Alternatively, the metallocene is first reacted with the aluminoxane to form a complex, and then the support is added to form the final catalyst. If an activated support (section 6.4.3) is required, the catalyst plant may need to do the support treatment step, which requires calcination equipment. Thus, supportation of the metallocene is likely to require different equipment than is used to prepare the metallocene. The reader can discern that a catalyst plant designed to prepare supported ZN catalysts is probably not equipped to prepare supported metallocene catalysts without substantial investment.

7.16   Concluding Remarks

From the above discussion it is clear that the manufacture of catalyst is an expensive and complex process. The high cost is justified by the large amount of polypropylene produced per pound of catalyst. In modern plants catalyst productivity exceeding 30,000 lb/lb is common, so the catalyst cost represents only a small fraction of the value of the polymer. The criticality of consistent catalyst quality drives daily operations at the plant. Even minor changes in the process may have subtle effects upon the catalyst that will be multiplied ten thousand fold in the final produced polymer. This is highly undesirable, and so once a manufacturing process is successfully fixed, changes are undertaken with the utmost caution and in careful consultation with the polymer manufacturing operations.

7.17   Questions

1. A particular catalyst plant finds that it has an impurity of an aromatic ester in both recovered titanium tetrachloride and also in the recovered hydrocarbon solvent that is used for washing the final catalyst. Assuming the level to be the same in both streams, which stream is likely to be of more immediate concern to the plant, and why?
2. A particular catalyst plant produces a supported catalyst at a fully burdened cost of $200/lb. The catalyst is applied in a polymerization plant that produces 40,000 lb/lb catalyst.

What is the cost of the supported catalyst per pound of polypropylene? How does this compare with the market value of commodity polypropylene? Is this the complete catalyst system cost?

3. A catalyst quality control laboratory has a laboratory polymerization test that produces 500 g of polymer per test. Good quality catalyst from the plant is known to produce 50,000 g polymer per gram of catalyst. How much catalyst can actually be tested in a bench polymerization? How does this impact the catalyst sampling protocol? List several factors which can strongly impact representative catalyst sampling.

4. Separate storage for virgin and recovered solvents is shown in the plant schematic in Figure 7.1. This is expensive. What factors could guide a decision on whether each virgin solvent and recovered solvent could be stored in a single tank? How would the production of several catalyst grades affect the decision making?

5. Compare two designs for titanium tetrachloride storage tanks: one design with gravity delivery from the bottom of the tank, one design with pump delivery from the top.

6. A catalyst manufacturer tests a new production lot and determines that performance is not meeting specifications. Further investigation reveals vanadium in the catalyst. What is the most likely source?

7. During the distillative recovery of spent $TiCl_4$ the heavy residues from the distillation column intermittently plug the discharge pump and line. What options might the plant consider to alleviate this problem.

References

1. For a discussion of some of the interactions during scale up, see *The Polymerization Catalyst Handbook*, R. Hoff and R.T. Masters, editors, John Wiley and Sons, 2010, chapter 5.

2. One notable exception to this is Toho Titanium, which only produces catalysts.

3. For a recent study on dealcoholation of a support see E.J. Dil, S. Pourmadian, M. Vatankhah, F.A. Taromi, *Polymer Bulletin*, 64, 445–457, 2010.

4. Detailed information on storage design and handling is found in Titanium Dioxide Manufacturers Association, "Safety Advice for Storage and Handling of Titanium Tetrachloride", 7th edition, June, 2009; http://www.cefic.org/files/Downloads/TiCl4-Safety-Advice-7th-Edition-June-2009-PA.pdf

5. S. Huffer, M. Kersting, F. Langhauser, R.A. Werner, S. Seelert, P. Muller, J. Kerth, US Patent No. 5773516.

6. D.D. Klendworth, F.W. Spaether, US Patent No. 7071137.

7. Z. Zhu, M. Chang, US Patent No. 7135531.

8. *Polypropylene Handbook*, 2nd Edition, edited by Nello Pasquini, Hanser Publishers, 2005, p 27.

9. Y. Yang, H. Du, Z. Li, Z. Wang, Z. Tan, K. Zhang, X. Xia, T. Li, X. Wang, T. Zhang, W. Chen, X. Zheng, US Patent Application 2006/0003888A1.

10. Y. Gulevich, I. Comurati, A. Cristofori, T. Dall"Occo, G. Morini, F. Piemontesi, G. Vitale, US Patent Application 2007/0282147 A1.

11. M. Ferraris, F. Rosati, S. Parodi, E. Giannetti, G. Motroni, E. Albizatti, US Patent No.4399054;

12. S. Huffer, M. Kersting, F. Langhauser, R.A. Werner, S. Seelert, P. Muller, J. Kuerth, US Patent No. 5773516.

13. U.S. Tanese,T.Sadashima, US Patent Application 2009/0197762 A1.

14. M. Gao, L. Xie, X. Wang, S. Zhao, J. Ma, H. Liu, T. Li, Z. Sun, US Patent No. 7323431.

15. M. Grun, K. Heyne, C. Mollenkopf, S. Lee, R. Uhrhammer, US Patent No. 7767772

16. Z. Zhu, M. Chang, US Patent No. 7135531.

17. M.T. Zoeckler, B.E. Wagner, S.C. Kao, U.S. Patent No. 7348383.

18. For a discussion of drying flammable slurries see GEA Processing Engineering web page: http://www.niroinc.com/technologies/special_cases.asp

19. *Handbook of Transition Metal Polymerization Catalysts*, R. Hoff and R.T. Mathers, editors, John Wiley and Sons, Inc., p 110, 2010.

20. N.F. Brockmeier, G.G. Arzoumanidis, N.M. Karayannis, N. Stein, *Chemical Engineering*, pp. 90–95, Sept. 1996.

21. J.T. Rasanen, "Optimisation of the Recovery Section of a Polyolefin Catalyst Manufacturing Process", Masters Thesis, Chalmers University, 2010.

22. Other methods, such as precipitation of impurities from solvents, have been reported. See for example, S.A. Cohen, J. A. Lee, D.B. Manley, US Patent No. 4914257; C.A. Drake, B. Loffer, US Patent No. 4588704.

23. See M.A. Nijenhuis, E.A.J.W. Kuppen, US Patent No. 7045480 and references therein.

24. For example, an aromatic ester might decompose in the solvent recovery system to form an aromatic acyl halide: $RCOOR' + TiCl_4 \rightarrow RCOCl + R'OTiCl_3$.

25. I. Pentti, P. Leskinen, US Patent No. 5641721.

26. Dusts will appear as a smoke. See Figure 9.5.

27. Metal Alkyls and Their Solutions: Burning Properties, Technical Bulletin, AkzoNobel, August 2008; http://www.akzonobel.com/hpmo/system/images/AkzoNobel_Metal_Alkyls_and_their_Solutions_Burning_Properties_ma_glo_eng_tb_tcm36–14982.pdf

28. Control of Metal Alkyl Fires, Technical Bulletin, AkzoNobel, August 2008; http://www.akzonobel.com/hpmo/system/images/AkzoNobel_Control_of_metal_alkyl_fires_ma_glo_eng_tb_tcm36–14977.pdf

29. M. Rigo, P. Canu, L. Angelin, G. Della Valle, *Ind. Eng., Chem. Res.*, vol. 37, pp. 1189–1195, 1998.

30. Typical specifications include limits on V, Fe, Mn, Ni, and Cu in the range of 1–10 ppm. See for example: http://www.toho-titanium.co.jp/en/products/ticl4_en.html

31. US Court of Appeals, Seventh Circuit INEOS POLYMERS INCORPORATED, Plaintiff-Appellant, v. BASF CATALYSTS and BASF Aktiengesellsch, Defendants-Appellees.No. 08–1359.

32. Olefin Polymerization Catalysts, Evonik Industries AG, http://catalysts.evonik.com/sites/dc/Downloadcenter/Evonik/Product/Catalysts/Brochures/425_Document.pdf

33. General Catalog Sud Chemie Catalysts, Sud-Chemie AG, 2007.

8

An Overview of Industrial Polypropylene Processes

8.1   Introduction

Principal process technologies employed in the manufacture of polypropylene are commonly known in the industry by a variety of descriptive names:

- slurry (aka "suspension")
- "loop slurry"
- bulk (aka "liquid pool")
- gas phase
- solution

Figure 8.1 shows technology platforms used today in the manufacture of polypropylene. LyondellBasell's Spheripol and the later generation Spherizone processes (discussed in section 8.7) are the most widely used. Table 8.1 summarizes key features of the industrial processes mentioned above. Typical plants have capacities of 200,000–300,000 mT/y.

Several modern technologies employ a cascade of processes. Most claim to be capable of producing homopolymer (HP), random

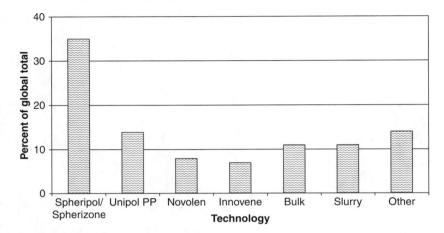

Figure 8.1 Polypropylene by technology platform. Source: G. Mai, International Conference on Polyolefins, Society of Plastics Engineers, Houston, TX, February 22–25, 2009.

copolymers (RACO) and heterophasic copolymers (HECO, aka "impact copolymers" or "ICP"). Note that the EPR component of HECO is typically produced in the final reactor in the series and under different conditions to generate the rubbery phase that imparts the excellent impact resistance of these products. HP and RACO may be produced in the same reactor and together account for more than 80% of the total global polypropylene market (see Figure 1.3). If a manufacturer elects not to participate in the HECO segment of the market, the downstream reactor and associated equipment become unnecessary and could be omitted to lower initial capital investment.

All processes must be capable of removing the considerable heat of polymerization of propylene. (The heat of polymerization of propylene is about 20 kcal/mole, slightly lower than the 22–26 kcal/mole values cited for ethylene.) This is achieved by various means using excess propylene monomer or an external source as cooling fluids. Except for the solution process, temperatures for propylene polymerization are typically controlled between about 60 and 80°C. Pressures are usually controlled below about 550 psig (~3.4 MPa).

As previously noted, Ziegler-Natta catalysts (discussed in Chapters 3 and 4) are used to manufacture more than 97% of the total global production of polypropylene as of this writing. Single site catalysts account for the balance. Modern process technologies claim to be capable of employing either Ziegler-Natta or single site catalysts.

Table 8.1 Features of industrial polypropylene processes.

Process Type	Typical Solvent	Typical Temperature Range (°C)	Typical Pressure Range (psig)	Early Practioners	Process Description
slurry (aka "suspension")	hexane	50–80	100–400	Montecatini, Hercules, Hoechst	Polymer precipitates to form a slurry; largely replaced by bulk and gas phase processes.
"loop slurry"	none (propylene)	70–90	200–400	Phillips (now Chevron Phillips)	Polymer precipitates in propylene to form a suspension
bulk (aka "liquid pool")	none (propylene)	45–80	250–550	Dart, El Paso, Sumitomo	Polymer precipitates in bulk ("liquid pool") propylene to form a suspension
gas phase	none	60–80	200–600	BASF, Amoco/Chisso, Union Carbide	Granular polymer formed in gas phase fluidized bed, horizontal stirred bed or a vertical stirred bed.
solution	hexane	≥140	300–500	Texas Eastman	Least used process for industrial polypropylene. Today used primarily to produce atactic polymer.

The preferred cocatalyst used with early Ziegler-Natta catalysts (1st and 2nd generations, see Chapter 4) was diethylaluminum chloride. DEAC was readily available at acceptable cost and performed well with early generation Ziegler-Natta catalysts. As supported catalysts (3rd generation) began to emerge in the 1970s, the cocatalyst that provided the best combination of cost and performance was triethylaluminum. Later, improved supported catalysts (4th and 5th generation) were introduced and also performed best with triethylaluminum. Occurring mostly over the period from the mid-1970s to the late-1980s TEAL displaced diethylaluminum chloride as the largest volume aluminum alkyl in the global market. This transition coincided with modern supported catalysts gradually supplanting early generation catalysts.

The most widely used cocatalysts for single site catalysts are methylaluminoxanes. However, SSC systems that require no cocatalyst have been recently developed, including "activated supports" (previously discussed in Chapter 6). However, please recall that SSC account for less than 3% of the total global production of polypropylene, as of this writing.

Most of the early processes (late 1950s-early 1960s) were slurry (suspension) processes where polymerization took place in an aliphatic hydrocarbon such as hexane [1]. A typical flowsheet for the slurry process as practiced by Hercules beginning in 1957 is reproduced in Figure 8.2. In the obsolete Hercules process, polypropylene precipitated as formed resulting in a slurry in hydrocarbon. The "loop slurry" process was developed later by Phillips Petrochemical (now Chevron Phillips Chemical) and is similar in that it produces a slurry, but uses a specially designed pipe reactor originally developed by Phillips in the 1950s for polyethylene. It was adapted to polypropylene in the early 1960s and modified, improved versions are still in use today. Further, the "loop slurry" process is typically conducted in liquid propylene and in that regard is similar to the bulk process. These early processes employed first generation catalysts (discussed in section 3.4 and Chapter 4) which were low in activity and produced significant amounts of atactic polymer. Consequently, post-reactor treatments were necessary to remove catalyst residues and to remove atactic polymer. Hogan and Banks [1] and Vandenberg and Repka [2] described these early processes.

More efficient processes began to appear in the mid- and late-1960s and accompanied the emergence of catalysts with higher activity and improved stereocontrol. Second generation

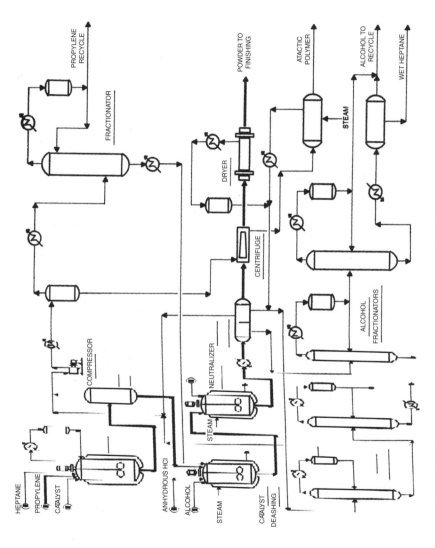

Figure 8.2 Flowsheet for early polypropylene slurry process as practiced by Hercules beginning in 1957. (P Spitzer, *Petrochemicals-The Rise of an Industry* (John Wiley & Sons) 337, 1988. Reprinted with permission of John Wiley & Sons).

(the so-called "precipitated" or "Solvay" catalysts, see the final paragraph in section 3.4 and section 4.4) and third generation supported catalysts (see section 3.5) increased both activity and isotactic index. Equipment for post-reactor treatment for removal of catalyst residues was obviated because transition metal residues were reduced to ppm levels. However, removal of atactic content was still required. Later, when even more highly stereoselective catalysts were developed, equipment for atactic removal became unnecessary. These developments greatly simplified processes and dramatically lowered capital investment and manufacturing costs.

A few additional observations are warranted to conclude the introduction to the various industrial processes for polypropylene. Historical names for the processes listed in Table 8.1 have become less meaningful in today's vernacular. Lines of distinction between processes have been blurred by use of multiple reactors operated under a range of conditions. For example, to produce impact copolymers, LyondellBasell's *Spherizone* process employs a "multi-zone circulating reactor" in combination with a fluidized bed reactor (see section 8.7). The reader should be aware that it is not uncommon for modern polypropylene technologies to use several reactors and these reactors may be operated either in series or in parallel. Moreover, each company may use its own trademarked name for the overall process. This makes it difficult to place some processes into a unique category. Nevertheless, the following sections deal with the common designations that have been used historically for industrial polypropylene process technologies. (Hybrid processes are surveyed in section 8.7.)

8.2   Slurry (Suspension) Processes

As noted above, the slurry (suspension) process was used by early manufacturers of polypropylene including Hercules in the USA and Montecatini and Hoechst in Europe. In the beginning, these processes were operated batchwise and employed an alkane diluent. Hexane was often used as the solvent for catalyst preparation (see section 5.4) and as the diluent in which polymerization was conducted. The cocatalyst of choice was diethylaluminum chloride because it was the metal alkyl that combined the best balance of availability, cost, activity, and polymer stereoregularity. Isotactic polymer precipitated as formed while much of the atactic polymer remained soluble in hexane. (However, additional treatment was still required to remove residual atactic polymer.) These processes were equipment-intensive and costly to operate (see Figure 8.2 and contrast that with the relative simplicity of a modern gas phase Unipol process in Figure 8.4).

As previously mentioned, early processes used inefficient 1st generation catalysts. Because these catalysts showed low activity and produced relatively large amounts of atactic polymer (>30%), resin required post-reactor treatment to remove catalyst residues and residual atactic polymer. Catalyst residues included a variety of complex titanium-, aluminum- and chlorine-containing by-products which will not be discussed here. Suffice it to mention that these residues could corrode downstream processing equipment and cause discoloration and decomposition of polymer.

Over the past few decades, the industry trend has been away from slurry (suspension) processes and that trend is expected to continue. Innovative reactor designs using bulk and gas phase processes (and combinations thereof) have largely rendered slurry processes passé. However, there are still a few plants around the world that employ the antiquated slurry process.

8.3   Bulk ("Liquid Pool") Process

Processes in which polymerization takes place in liquid propylene (functioning as both reactant and solvent) are known as "bulk" or "liquid pool" processes. Polymer precipitates as formed and, in that regard, bulk processes are similar to the slurry process. The first bulk process was developed in the 1960s by Dart Industries. (The Dart process is sometimes also called the Rexene, Rexall or El Paso bulk process.) In most cases, polymerization takes place in continuous stirred tank reactors or autoclaves, though Chevron Phillips Chemical uses a pipe reactor (see discussion of the CPChem "loop slurry" reactor in next section). Sumitomo also developed a bulk process and licensed the technology, most notably to Exxon.

8.4   "Loop Slurry" Process (Chevron Phillips Chemical)

Previously noted, the continuous process as practiced today by CP Chem could be considered both a slurry and a bulk process, since polymerization is carried out in a jacketed pipe reactor in rapidly circulating liquid propylene. As in the bulk process, propylene serves both as solvent and reactant. The CPChem process is shown schematically in Figure 8.3. A key distinction between the CPChem process and early bulk processes is the vessel in which the polymerization takes place. In the bulk (aka "liquid pool") process, continuous stirred tank reactors or autoclaves are most common, whereas, a series of pipe reactors (up to 8 "legs") are used in the CPChem process. However, Chevron Phillips calls their process the "loop slurry" process (*i.e.*, not a "loop bulk" process). The loop design provides the maximum surface area to remove the heat of polymerization. As the heat removal capacity increases, the throughput of the reactor can be increased.

8.5 Gas Phase Processes

Several gas phase processes have been developed over the past 40+ years. They have in common a gas phase propylene polymerization in the virtual absence of solvents or diluents, but they differ in reactor configurations, the catalysts and the manner in which the components are mixed. Though supported Ziegler-Natta catalysts are used almost exclusively, the catalysts are prepared *via* dissimilar schemes and differ significantly in

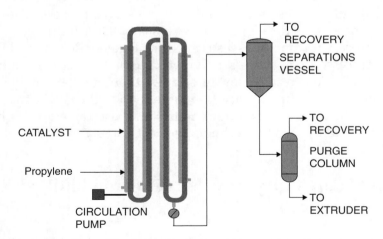

Figure 8.3 Schematic flowsheet for Chevron Phillips loop slurry process (Reproduced with permission of B. Beaulieu, Chevron Phillips Chemical).

composition. In gas phase processes it is unnecessary to remove atactic polymer, because modern highly stereospecific catalysts are used.

The Unipol gas phase process for polypropylene is an extension of the Union Carbide (now Dow Chemical) gas phase process for polyethylene originally developed in the late 1960s. In the early stages of development of the "UNIPOL PP" process, the Shell high activity catalyst ("SHAC") was used. More recently, 3rd and 4th generation SHAC catalysts are employed [3]. The UNIPOL PP process utilizes a vertical fluidized bed reactor and no mechanical agitation is required. The bulbous upper portion of the reactor has become almost iconic in the polyolefins industry. The bulged part of the reactor is designed to be a particle disengagement zone, allowing suspended particles to fall back into the fluidized bed. It is important to use a catalyst containing a minimum of both very large particles and very fine particles. The former produces large polymer agglomerates that are hard to fluidize, and the latter produces very small polymer particles that are more difficult to disengage from the circulating propylene.

The fluidized bed is maintained by introducing propylene gas through a distributor plate at the bottom of the reactor. The flow rate is very high and the propylene serves both as monomer and as cooling gas to remove the heat of polymerization. Excess propylene gas is cooled and returned to the reactor. Static buildup can be a problem. It arises from the friction developed between the flowing gas and the non-conducting polymer particles. When static charges build up the reactor wall can accumulate a polymer sheet, which can shut down the fluidized bed if it dislodges from the wall. Special additives are introduced in small amounts into the reactor to control the static charge.

Union Carbide/Dow started up the first UNIPOL PP plant in 1985. UNIPOL PP has been the most successful of the gas phase polypropylene technologies, having been licensed to more than 20 companies worldwide operating 46 UNIPOL PP lines, as of 2011 (R. Patel, personal communication, January, 2012). In a review article in 2005 [3], Engel claimed that Union Carbide/Dow had "over 120 polyolefin reactor lines worldwide" and "1700 reactor-years of safe operations" for UNIPOL process technology as of the end of 2003. Presumably, those statistics include figures for the UNIPOL process for polyethylene and have increased

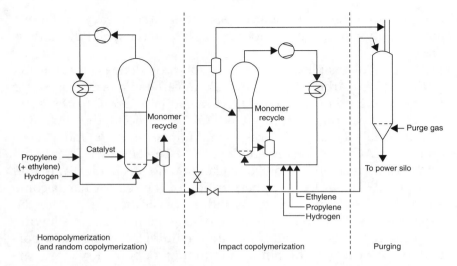

Figure 8.4 Simplified flow diagram for Unipol gas phase process from K Whitely, T Heggs, H Koch, R Maner and W Immel, *Ullman's Encyclopedia of Industrial Chemistry*, Wiley-VCH, Wenheim, 6th edition, Vol 28, 443, 2005. (Reprinted with permission of Wiley-VCH).

substantially since 2003. Figure 8.4 shows a schematic for the Unipol gas phase process.

The Lummus-Novolen gas phase process employs a vertical stirred bed. This process, originally developed by BASF in the late 1960s, employs proprietary helical agitators [4]. The Lummus-Novolen process is shown schematically in Figure 8.5. The reactor can operate more than ¾ full. The process temperature must be carefully controlled at several points in the reactor because the stirred bed is prone to developing hot spots. Because it is mechanically stirred it is less susceptible to wall fouling, which is a problem for fluidized bed gas phase reactors. Unlike fluidized bed gas phase processes, the amount of gas that is required to be recycled is much lower, because it only functions for heat removal, not mixing.

The Amoco/Chisso (now INEOS) technology is known today as the Innovene PP process and employs a horizontal stirred bed gas phase reactor. A second horizontal reactor is used for HECO. The reactor contains four connected compartments, each with a paddle agitator along a horizontal axis. The individual sections of the reactor give the process certain plug-flow characteristics. The process is illustrated in Figure 8.6 [5].

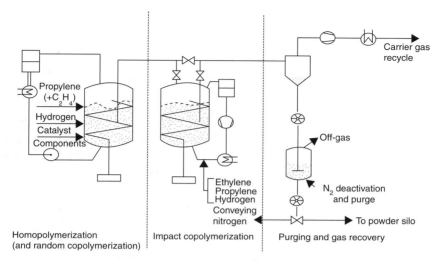

Figure 8.5 Simplified flow diagram for Lummus-Novolen (BASF) gas phase process from K Whitely, T Heggs, H Koch, R Maner and W Immel, *Ullman's Encyclopedia of Industrial Chemistry* (Wiley-VCH, Wenheim), 6th edition, Vol 28, 442, **2005**. (Reprinted with permission of Wiley-VCH).

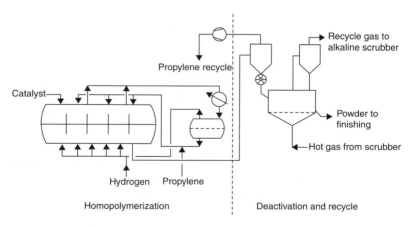

Figure 8.6 Simplified flow diagram for Amoco-Chisso (INEOS) gas phase process from K Whitely, T Heggs, H Koch, R Maner and W Immel, *Ullman's Encyclopedia of Industrial Chemistry,* Wiley-VCH, Wenheim, 6th edition, Vol 28, 443, **2005**. Does not depict gas phase reactor used for HECO downstream of homopolymerization reactor. (Reprinted with permission of Wiley-VCH).

8.6 Solution Process

The solution process is the least common method for industrial production of polypropylene and differs from the processes discussed above in that the operating temperature is >100°C [1, 6, 7]. Though judged to be obsolete in a 1986 review [1], the solution process is still practiced commercially for production of atactic polypropylene. Also, Dow has recently developed a solution process for production of specialty propylene-ethylene plastomers and elastomers called VERSIFY with densities in the range of 0.85–0.89 g/cc. Dow INSITE single site catalysts are used to produce VERSIFY copolymers which show PDI of 2–3. (For more on VERSIFY products see *www.dow.com/versify/about/*.) The total quantity of polypropylene produced *via* solution processes is not known but is believed to be quite small relative to stereoregular polypropylene.

8.7 Hybrid Processes

As noted above, LyondellBasell's Spheripol and Spherizone processes are the most widely used, together accounting for about ⅓ of the total global production of polypropylene (see Figure 8.1). However, neither can be placed exclusively in one of the historic categories discussed above. Both are hybrid processes. Each is capable of producing the complete range of polypropylene products (HP, RACO and HECO).

The Spheripol process employs two loop slurry reactors in combination with a gas phase reactor [8] and is shown schematically in Figure 8.7. Operating conditions in the loops are 60–80°C and 35–50 atm, and operating conditions in the gas phase reactor are up to 100°C and 15 atm. The Mitsui Chemicals Hypol II process is similar to the Spheripol process but employs the Mitsui high yield-high stereospecificity (HY-HS) catalysts. A miniature loop reactor is used to prepolymerize catalyst, which is then introduced into the main loops.

LyondellBasell's Spherizone process employs a "multizone circulating reactor" (MZCR) which is a specially designed loop reactor that polymerizes propylene under different reactor regimes [9]. The flow in the loops is maintained by specially designed pumps. For HECO, a gas phase reactor is used downstream of the MZCR.

A schematic of the Spherizone process is shown in Figure 8.8. As with the Phillips loop process there is a very large circulating inventory of flammable propylene.

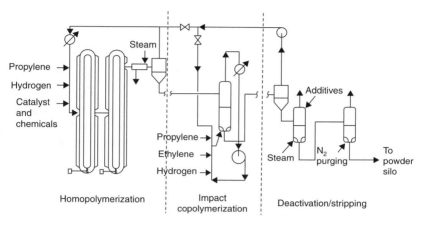

Figure 8.7 Simplified flow diagram for LyondellBasell Spheripol process from K Whitely, T Heggs, H Koch, R Maner and W Immel, *Ullman's Encyclopedia of Industrial Chemistry*, Wiley-VCH, Wenheim, 6th edition, Vol 28, 441, **2005**. (Reprinted with permission of Wiley-VCH).

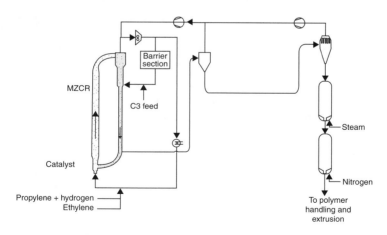

Figure 8.8 Simplified flow diagram for LyondellBasell Spherizone process from R. Lieberman, M Dorini, G Mei, R Rinaldi, G Penzo and GT Berg, *Kirk-Othmer Encyclopedia of Chemical Technology*, John Wiley & Sons, Inc., Vol 20, 541, **2006**; does not depict the gas phase reactor used for HECO downstream of the MZCR. (Reprinted with permission of John Wiley & Sons, Inc.)

Another hybrid is the Borstar process developed by Borealis. Like Spheripol, Borstar uses a prepolymerized catalyst and combines a loop slurry process with a gas phase reactor for HP and RACO [10]. The Borstar process uses a small loop reactor for prepolymerization of catalyst immediately prior to introduction to the large loop reactor. Unlike Spheripol, however, Borstar employs a second smaller gas phase reactor to produce HECO.

The Borstar process is unique in that it is capable of using supercritical propylene, operating above 91°C. Under these conditions, using special catalysts, productivity is enhanced. Under supercritical conditions the amount of hydrogen that can be added is unrestricted; bubbles do not form as they would in the propylene below the critical point. This permits the production of polymer grades with very high melt flow rates.

8.8 Kinetics and Reactivity Ratios

An important variable in polypropylene process technology for copolymers is the difference in reactivity between propylene and the comonomer, most often ethylene. In processes for production of propylene-ethylene copolymers (RACO and HECO), ethylene is always the more reactive olefin. Feed proportions used in commercial processes to achieve a desired copolymer composition are determined in consideration of what is called the reactivity ratio. Reactivity ratios are based on kinetics and their determinations commonly require measuring rate constants for several propagation reactions. Possible reactions that can occur in producing copolymers are summarized below. They may be classified as self-propagation or cross-propagation reactions:

- **Self-propagation:**

$$Met\text{-}CH_2\text{-}\overset{\overset{\displaystyle CH_3}{\displaystyle |}}{CH} \sim \; + \; CH_2 = CHCH_3 \quad \xrightarrow{\;k_{pp}\;} \quad Met\text{-}CH_2\text{-}\overset{\overset{\displaystyle CH_3}{\displaystyle |}}{CH}\text{-}CH_2\text{-}\overset{\overset{\displaystyle CH_3}{\displaystyle |}}{CH} \sim \quad (8.1)$$

- **Cross-propagation:**

$$Met\text{-}CH_2\text{-}\overset{\overset{\displaystyle CH_3}{\displaystyle |}}{CH} \sim \; + \; CH_2 = CH_2 \quad \xrightarrow{\;k_{pe}\;} \quad Met\text{-}CH_2\text{-}CH_2\text{-}CH_2\text{-}\overset{\overset{\displaystyle CH_3}{\displaystyle |}}{CH} \sim \quad (8.2)$$

- **Self-propagation:**

$$Met\text{-}CH_2\text{-}CH_2\sim + \ CH_2 = CH_2 \ \xrightarrow{\quad k_{ee} \quad} \ Met\text{-}CH_2\text{-}CH_2\text{-}CH_2\text{-}CH_2\sim \qquad (8.3)$$

- **Cross-propagation:**

$$Met\text{-}CH_2\text{-}CH_2\sim + \ CH_2 = CHCH_3 \ \xrightarrow{\quad k_{ep} \quad} \ Met\text{-}CH_2\text{-}\overset{\overset{\displaystyle CH_3}{|}}{C}H\text{-}CH_2\text{-}CH_2\sim \qquad (8.4)$$

where:

- Met = metal active center (usually Ti)
- k_{pp} = rate constant for a terminal propylene group adding another propylene unit (propylene self-propagation as in eq 8.1)
- k_{pe} = rate constant for a terminal propylene group adding an ethylene unit (propylene-ethylene cross-propagation as in eq 8.2)
- k_{ee} = rate constant for a terminal ethylene group adding another ethylene unit (ethylene self-propagation as in eq 8.3)
- k_{ep} = rate constant for a terminal ethylene group adding a propylene unit (ethylene-propylene cross-propagation as in eq 8.4)

Reactivity ratios are then expressed as follows:

$$r_1 = k_{pp}/k_{pe}$$
$$r_2 = k_{ee}/k_{ep}$$

The product of r_1 and r_2 gives information about the expected composition of the copolymer. For example, if $r_1 \sim r_2 \sim 1$, monomer and comonomer will be incorporated in near equal proportions and the polymer will be a truly random copolymer.

Determinations of reactivity ratios are not straightforward and require sophisticated experimentation. Fortunately, reactivity ratios for many Ziegler-Natta catalyst systems have been published. Derivations are considered beyond the purview of this text. However, more details on derivations of reactivity ratios have been provided by Stevens [11], Krentsel, *et al.* [12] and Kissin [13].

8.9 Emergency Stoppage of Polymerization

Typically, modern industrial processes for polypropylene are equipped with sophisticated sensors, flow meters and computer controls. Reactors are closely monitored for upsets. Decompositions of the type that occasionally occur in high pressure free radical ethylene polymerizations [14] are not observed with propylene. Though rare, however, it is possible for runaway propylene polymerizations to occur. Because of the high heat of polymerization of propylene, uncontrolled polymerization can result in excessive internal temperatures that can melt the polymer. As a safety measure, many industrial polypropylene processes incorporate in their design the capability to terminate polymerizations quickly (*i.e.*, to "shortstop" the process). The most common technique is to introduce a poison that destroys or blocks active centers of the catalyst and immediately shuts down the polymerization. Gas phase processes are more susceptible to uncontrolled exotherms than liquid phase processes because of lower heat removal capacity. An uncontrolled exotherm can fuse the polymer particles into a giant mass, requiring manual cleaning of the reactor with chainsaws, which is a hazardous, laborious and expensive process.

Reagents that destroy the catalyst include protic compounds such as alcohols. However, any compound that contains a labile proton will cleave the metal-carbon bond in the catalyst, thereby irreversibly destroying the active center for polymerization.

Some reagents are capable of stopping the polymerization by occupying coordination sites on the active centers of the transition metal catalyst (see eq 4 in Chapter 3). This effectively blocks the coordination site to incoming propylene, thereby terminating polymerization. The most widely used compound in industry is carbon monoxide. CO rapidly poisons the catalyst and stops the polymerization. However, CO is a *reversible* poison. Once the process is under control again, CO can be removed and polymerization safely resumed. Another advantage of carbon monoxide is that it is unreactive with aluminum alkyls and therefore will not consume cocatalyst as would CO_2.

Isotopically labeled CO has been used for investigations of the mechanism of polymerization [15].

8.10 Questions

1. What process and type of catalyst and cocatalyst were most used in early manufacturing of PP?
2. Why were early PP plants much more equipment-intensive than modern plants?
3. Why did TEAL displace DEAC as the most widely used aluminum alkyl in PP manufacture?
4. A modern polypropylene plant has a capacity of 550,000,000 lb/y. The plant employs a catalyst that contains 3.10 % titanium and has an activity of 40,000 lb of PP per lb of catalyst. Assuming that the plant runs at full capacity and the process requires an Al/Ti molar ratio of 80, how many pounds of TEAL would be required for one year of operation?
5. Assuming that all of the aluminum and titanium are retained in the polymer in problem 4, what will be the residual content in ppm of Ti and Al in the polymer?
6. How does carbon monoxide function in terminating polymerization?

References

1. JP Hogan and RL Banks, *History of Polyolefins*, (R. Seymour and T. Cheng, editors), D. Reidel Publishing Co., Dordrecht, Holland, 111, 1985.
2. EJ Vandenberg and BC Repka, *High Polymers*, (ed. CE Schildknecht and I Skeist), John Wiley & Sons, *29*, 369, 1977.
3. BR Engle, *Handbook of Petrochemicals Production Processes*, (RA Meyers, ed.), McGraw-Hill, NY, 16.62, 2005.
4. K. Whitely, T Heggs, H Koch, R Maner and W Immel, *Ullman's Encyclopedia of Industrial Chemistry*, 6th edition, Wiley VCH Verlag GmbH, Weinham, Vol 28, 441.
5. ibid., p 443; see also WD Stephens, *International Conference on Polyolefins*, Society of Plastics Engineers, Houston, TX, February, 2011.
6. D Bigiavi, M Covezzi, M Dorini, R LeNoir, R Lieberman, D Malucelli, G. Mei and G. Penzo, *Polypropylene Handbook*, (N. Pasquini, editor), 2nd edition, Hanser, 365, 2005.
7. K. Whitely, T Heggs, H Koch, R Maner and W Immel, *Ullman's Encyclopedia of Industrial Chemistry*, 6th edition, Wiley VCH Verlag GmbH, Weinham, Vol 28, 440.
8. J Severn and RL Jones, Jr, *Handbook of Transition Metal Polymerization Catalysts*, R Hoff and R Mathers (editors), Wiley, 165, 2010.

9. R Rinaldi and G ten Berge, *Handbook of Petrochemicals Production Processes,* (RA Meyers, ed.), McGraw-Hill, NY, 16.21, 2005.

10. J Kivela, H Grande and T Korvenoja, *Handbook of Petrochemicals Product Processes,* (RA Meyers, ed.), McGraw-Hill, NY, 16.41, 2005.

11. M Stevens, *Polymer Chemistry,* 3rd ed., Oxford University Press, New York, p 194 (1999).

12. B Krentsel, Y Kissin, V Kleiner and L Stotskya, *Polymers and Copolymers of Higher a-Olefins,* Hanser-Gardiner Publications, Cincinnati, OH, p. 244 (1997).

13. YV Kissin, *Alkene Polymerization Reactions with Transition Metal Catalysts,* Elsevier, Amsterdam, p. 190 (2008).

14. DB Malpass, *Introduction to Industrial Polyethylene,* (published jointly by Scrivener Publishing, LLC and John Wiley & Sons), p. 29, 2010.

15. J Severn and RL Jones, Jr, *Handbook of Transition Metal Polymerization Catalysts,* R Hoff and R Mathers (editors), Wiley, p 185, 2010; see also YV Kissin, *Alkene Polymerization Reactions with Transition Metal Catalysts,* Elsevier, Amsterdam, p. 491 (2008).

9

Laboratory Catalyst Synthesis

9.1   Introduction

The purpose of this Chapter is to provide the reader with an appreciation of Ziegler-Natta catalyst synthesis by providing detailed descriptions of two representative industrial catalyst preparations. Although there are many hundreds of examples to be found in the literature, almost all patent examples and most journal examples omit many experimental details on equipment set up and handling that are useful to the laboratory researcher. The syntheses in this chapter depict examples of two types of catalyst: a supported catalyst incorporating di-*n*-butylphthalate (sometimes called a 4th generation catalyst), and a precipitated $TiCl_3$ catalyst (sometimes called a second generation or Solvay catalyst). The latter gives a catalyst which, under industrial polymerization condition, produces perhaps 5 kg polymer per gram catalyst of broad molecular weight polypropylene, whereas the former produces about 50 kg of polymer per gram of catalyst of a more narrow molecular weight polypropylene. Similar activities can be achieved under laboratory conditions.

By examining the details of these two syntheses the reader will gain an appreciation for the length, complexity, and raw material usage involved in the preparation of industrially relevant catalysts. It is anticipated that the descriptions will provide enough detail to guide the skilled organic chemist through the preparations. A good way of visualizing the syntheses is by flow charts. These are shown in Figure 9.1 and Figure 9.2. Reagents input are on the left hand side, and effluents are shown on the right hand side of each diagram. Boxes represent the reaction steps. The Figures demonstrate that catalyst synthesis requires a large number of steps with very significant effluent streams.

9.2   General Synthesis Requirements

First and foremost, the chemist requires a working knowledge of the techniques of safe chemical handling under anaerobic conditions. These syntheses should be performed by a competent synthetic chemist who has acquired experience from handling Grignard reagents, alkali metals, or other air-sensitive reagents. In an academic laboratory, advanced undergraduate students may be capable of executing the syntheses, provided that there is close supervision and careful attention to safety.

Ziegler-Natta catalyst synthesis in the laboratory requires typical organic synthesis glassware and a good fume hood. The general philosophy adopted in this chapter is to do all operations in the mother vessel, and to avoid all Schlenk operations that require lifting or tilting glassware [1]. Although such Schlenk techniques can be elegant, they introduce unnecessary safety hazards. Doing the chemistry in stationary equipment simplifies operations and reduces hazards from the presence of hot titanium tetrachloride, aluminum alkyls and organic solvents.

The scale of the syntheses will produce about 10 g of catalyst. This is sufficient for characterization and polymerization tests.

9.3   Equipment Requirements

For the following syntheses the facility and equipment requirements include the items listed below. Diagrams are shown in

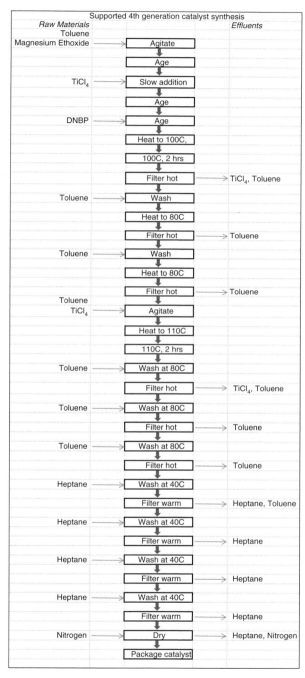

Figure 9.1 Supported 4th generation catalyst synthesis flowchart.

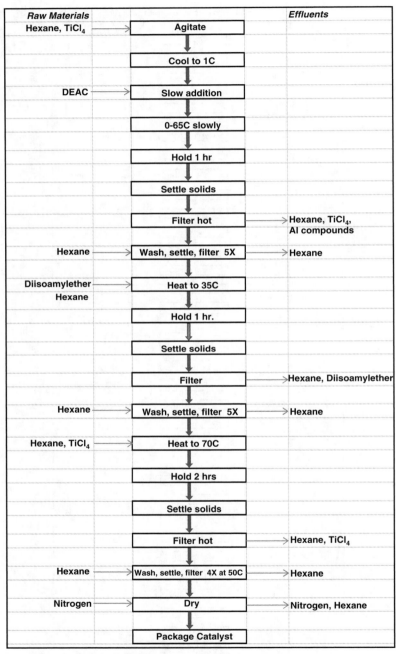

Figure 9.2 Precipitated TiCl$_3$ synthesis flowchart.

Figures 9.3 and 9.4. Glassware may be sourced through various scientific glassware vendors [2]. Other items are available from laboratory supply houses [3].

Equipment:

- Well-ventilated hood, at least four feet in length, with moveable window sashes
- Safety shower and eyewash located nearby
- Supply of powdered clay or vermiculite for spills
- Hood lattice or other sturdy support system
- Adjustable clamps
- 5-neck, 250-mL and 500-mL round bottom glass flasks with 14/20 or 24/40 standard taper joints [4]
- 50-mL and 250-mL pressure equalized dropping funnels [5]
- Nitrogen feed line with low pressure regulator, set at 1–4 psig
- Flexible Teflon transfer tubing
- Rubber tubing
- Double beveled metal cannula
- Coarse frit filter tube [6] and threaded adapter assembly [7]
- Filtrate waste receiver bottle
- Laboratory jacks
- Rubber septa, to fit dropping funnel and filtrate waste receiver
- Disposable 2.0-mL and 5.0-mL syringes and 16-gauge needles
- Vacuum pump or other vacuum source and trap
- 250-mL titanium tetrachloride reservoir and 500-mL solvent reservoir bottles
- Mechanical stirring assembly with single Teflon blade, flexible shaft
- Variable speed motor
- Tachometer
- Thermocouple and temperature controller [8]
- Oil bath with heating coil or electric heating mantle
- Flame-retardant lab coat, safety glasses, face shield
- Neoprene gloves and disposable nitrile gloves
- Type B and type D fire extinguishers [9]
- Acid vapor suppression extinguisher [10]

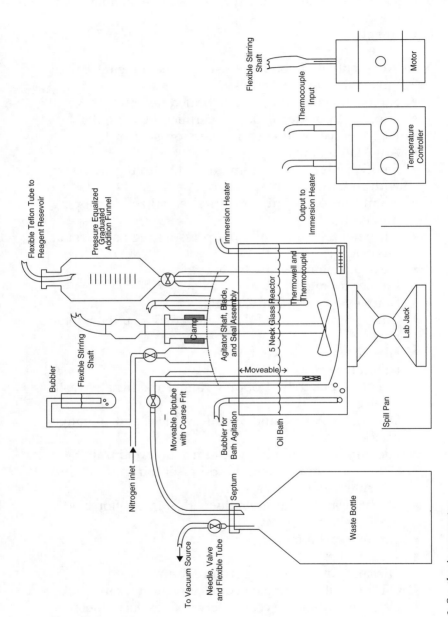

Figure 9.3 Synthesis apparatus.

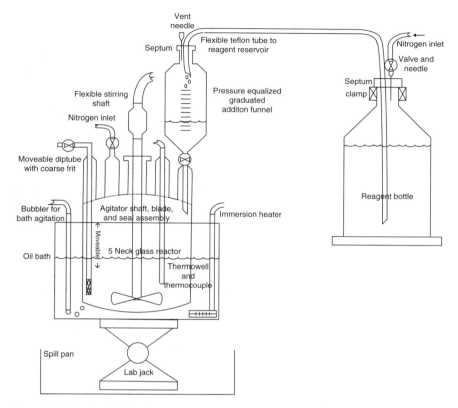

Figure 9.4 Reagent transfer apparatus.

Alternatively, more sophisticated laboratory glassware can be utilized: a two-piece resin flask, or jacketed reactor with an external oil circulating bath, and precision pumps for metering liquid deliveries.

9.4   Synthesis Schedule

The preparation of Ziegler-Natta catalysts is time-consuming and will make for long work-days. For the laboratory researcher, this means identifying convenient stopping points where chemistry of the process is not unduly affected. Recommended stopping points are identified in the following syntheses by "(STOP)".

9.5 Handling TiCl$_4$

Titanium tetrachloride, TiCl$_4$, is a hazardous material and many skilled organic chemists may never have handled this material. It is a free-flowing (non-viscous) liquid that, on contact with moisture in air, releases hydrochloric acid vapors and voluminous titanium oxychloride smoke. See Figure 9.5. Both cause chemical and thermal burns on contact with skin or eyes, and respiratory damage on inhalation. Thus, it is imperative that TiCl$_4$ be handled in a fume hood. In case of skin or eye contact, or of inhalation, the basic treatment is to remove to fresh air, flush affected areas with large amounts of water and seek immediate medical attention. Detailed information on handling, hazards, and first aid is provided in the material safety data sheet that accompanies the product, and users must familiarize themselves prior to use [11]. The TWA is 0.5 mg/m^3. The laboratory usually receives TiCl$_4$ in quart bottles or steel containers of a few gallons capacity. Transfer of TiCl$_4$ into the experimental

Figure 9.5 Titanium tetrachloride smoke. when TiCl$_4$ is exposed to the atmosphere there is immediate fuming. The smoke is composed of hydrated titanium oxychlorides and hydrogen chloride.

apparatus is best done by positive displacement using 1–2 psig nitrogen pressure and a cannula or flexible transfer tube. A typical set up is shown in Figure 9.4. All transfers and handling must be done in a well-ventilated hood. A face shield is recommended during transfers. A safety shower and safety eye wash must be located nearby. Small spills are handled by covering with powdered clay or vermiculite and neutralizing cautiously with soda ash in water. Large spills are treated with acid suppression foam to reduce HCl fumes and subsequently can be treated with solid absorbents.

9.6 Handling Diethylaluminum Chloride

Neat diethylaluminum chloride, DEAC, is a pyrophoric liquid; on contact with air it will burst into flames. Although there are methods to handle the neat liquid in the laboratory safely, it is still preferable to purchase it as a 10 wt% solution in an inert hydrocarbon, where its concentration is below the NPL (non-pyrophoric limit) [12]. This may reduce, but does not eliminate, handling risks. Accordingly, the solution must be transferred by a cannula or syringe with proper safety equipment; contact with oxygen or moisture instantly results in decomposition and one must still respect the potential for ignition. A face shield and flame-retardant Nomex laboratory coat is recommended during handling [13]. Detailed handling instructions accompany the sample bottle which usually comes with a secured septum seal. In the event of a small diethylaluminum chloride fire confined inside a hood, a type D fire extinguisher can be used to smother the flames, recognizing there is a danger of reignition. *DO NOT USE WATER, because explosive reaction will occur.* Large quantities of neat DEAC should not be handled in laboratory glassware. Also, carbon dioxide extinguisher will be ineffective and should not be used. If there are no reservoirs of flammables under threat, the best response may be to step back and allow the fire to burn itself out.

9.7 Spent Liquids

The synthetic procedures produce far more spent liquids than solid catalyst product. These must be collected, stored, and disposed of in a responsible and safe manner. The waste will consistent largely of liquid hydrocarbons with precipitated titanium oxyhalides. As such, they are flammable and acid-corrosive materials that react with water to evolve heat and acid fumes. In case of waste liquids containing alkylaluminum compounds, reaction with air will produce ethane which may build pressure within the waste container. Waste containers should be kept in a hood and properly labeled to prevent addition of other, incompatible materials.

9.8   Synthetic Procedure for Fourth Generation Supported Catalyst

9.8.1   Introduction

This synthesis is adapted from an example in the patent literature [14]. It describes the preparation of a fourth generation high activity, high stereoselectivity, titanium catalyst supported on magnesium chloride with a diester internal donor (see chapter 4.2.4). The magnesium chloride support is formed by a controlled halogenation with $TiCl_4$, and modified by contact with the internal donor, di-n-butylphthalate. The support is then extracted repeatedly with a blend of toluene and titanium tetrachloride which completes the halogenation of the support and increases the catalytic activity. Fifth generation catalysts, which use substituted 1,3 dimethoxypropanes internal donors or substituted succinates, follow similar preparative methods. However, those internal donors are not as readily available as the phthalate esters [15, 16].

9.8.2   Procedure

Assemble the apparatus in Figure 9.3 with a 500-mL reactor, except for the oil bath. Thoroughly purge the apparatus with dry nitrogen. Charge 120 mL dry toluene to the 500-mL liter glass reactor via the dropping funnel. Add 15.0 g magnesium ethoxide (ex: Catylen® S 100) using a wide-mouth funnel. This should be done quickly because $Mg(OEt)_2$ is hygroscopic. After checking that the bottom valve is closed, add 25 mL $TiCl_4$ to the 50-mL dropping funnel by a cannula. See Figure 9.4. Add a low viscosity silicone oil bath around the glass reactor and chill the bath with dry ice chips, agitating at ~ 300 rpm until the reactor reaches ~10°C. Add the $TiCl_4$ from the dropping funnel at a rate that maintains the temperature below

10°C, and then maintain for one hour at 10°C. Raise and lower the bath as needed to maintain the temperature. Under a gentle nitrogen flow, with a disposable syringe and needle add 2.0 mL of di-*n*-butyl phthalate through the septum on the dropping funnel. Remove the cooling bath and replace with a heating oil bath. Heat to 100°C and maintain for two hours. Stop agitation and allow the solids to settle. Lower the filter dip tube into the hot liquid and, by applying slight negative pressure to the receiver (Figure 9.3), remove as much supernatant liquid as possible. (STOP)

Add 120 mL toluene by cannula to the solids in the reactor, begin agitation, and heat to 80°C. Maintain for 10 minutes and remove supernatant liquid as before. Repeat this washing/decantation sequence one more time. Add 120 mL toluene and then the remaining 25 mL $TiCl_4$ in the dropping funnel to the reactor. Heat to 110°C at 300 rpm and maintain for 2 hours. Stop the agitation, filter as before, and add 120 mL toluene into the reactor by cannula. (STOP)

Begin agitation, heat to 80°C, maintain for 10 minutes, settle, and decant. Repeat this toluene treatment two more times. Add 130 mL dry heptane to the reactor by cannula, warm to 40°C and maintain for 10 minutes with agitation, and then filter. Repeat this washing and filtration 3 times. Dry the final catalyst in the flask by a gentle stream of dry, oxygen-free nitrogen stream at ~ 40°C until a free flowing powder is obtained. Collect the catalyst in an inert atmosphere glove box. The yield will be approximately 15 gm of catalyst.

9.9 Synthetic Procedure for Second Generation Precipitated $TiCl_3$ Catalyst

9.9.1 Introduction

This example is derived from the patent literature [17] of Solvay Polyolefins, which pioneered work on this type of catalyst. Historically, it was a big advance over previous milled versions of titanium tetrachloride, both in activity and stereospecificity. Although since surpassed by the $MgCl_2$-supported catalysts in performance, this synthesis is a good example of particle morphology and size-distribution control that is readily achieved in the laboratory. The synthesis consists of the reduction of titanium tetrachloride with diethylaluminum chloride to beta (β) titanium trichloride from a hexane solution under controlled agitation. This is followed by extraction of coprecipitated aluminum salts with a combination of $TiCl_4$ and diisoamylether and concurrent conversion of the titanium trichloride from the β (beta) crystalline form into the more active δ (delta) form.

9.9.2   Procedure

Assemble the apparatus of Figure 9.3 using a 500-mL liter glass reactor. Thoroughly purge the apparatus with dry nitrogen. Remove flammable solvents from the hood and assemble the set up in Figure 9.4 for the transfer of DEAC solution into the apparatus. After first checking that the bottom valve is closed on the dropping funnel, transfer ~166 g of 10 wt% DEAC solution in hexane to the 250 mL graduated dropping funnel by cannula using a slight nitrogen pressure to motivate the solution. Remove the DEAC bottle to storage. Prepare a 0.5-liter reservoir of deoxygenated dry hexane by sparging with nitrogen for 20 minutes in the hood. Add 60.0 mL hexane and 15.0 mL $TiCl_4$ (density = 1.73 g/mL, 26.0 g) into the glass reactor by syringe or cannula. Begin gentle agitation at 160 rpm and purge the reactor for 5 minutes with nitrogen. During this time cool the reactor with an insulated oil bath cooled with dry ice chips. Continue cooling until the reactor contents are at 1°C. Begin adding the DEAC solution at 0.8 mL/minute while maintaining the temperature at 1°C by raising/lowering the cooling bath as needed. Reduce/suspend the DEAC addition as needed to maintain the temperature. The addition will take about 4 hours. Maintain the temperature at 0°C for an additional 15 minutes, continue stirring and remove the cooling bath. Using the oil bath, bring the reactor to room temperature over one hour, then raise the reactor to 65°C over 1.5 hours and maintain that temperature for one more hour. Stop agitation and allow the solids to settle completely. Lower the filter tube slowly into the mixture and remove as much hot solvent as possible by applying a slight negative pressure to the receiver. Add ~50 mL hexane and wash the brown solids for 5 minutes with stirring, then decant the supernatant with the filter tube. Repeat the washing step four more times, then add 160 ml hexane. (STOP)

Add 26 mL degassed diisoamylether into the reactor *via* the dropping funnel. Heat to 35°C with stirring at 200 rpm and maintain at 35°C for one hour. Stop the agitation and allow solids to settle to bottom of flask. Then the supernatant liquid is removed with the filter tube as before. Add ~50 mL hexane and wash the brown solids for 5 minutes with stirring, then decant with the filter tube. Repeat the washing step four more times. Add 50 mL degassed hexane and 35 mL $TiCl_4$. These can be measured by performing the addition through the dropping funnel or by using a syringe. Resume stirring at 100–150 rpm and heat to 70°C. Maintain at 70°C for two hours, then allow the solids to settle and decant while hot, as before. Add 50 mL hexane, heat to 50°C with gentle stirring for 10 minutes, allow the solids to settle and decant the hot solvent through the filter tube. Repeat the washes three times. Dry the wet solids under a flowing stream of nitrogen, and collect the free flowing violet solid catalyst in an inert atmosphere glove box. The yield will be about 25 g.

9.10   Catalyst Analysis

The titanium content of a catalyst is conveniently determined by dissolution in aqueous sulfuric acid and H_2O_2 and absorbance measurement in the 410–450 μ (micron) range of the visible spectrum. The measurement is then referenced against a standard calibration curve. Other methods include atomic absorbance and inductively coupled plasma emission spectrometry [18]. Organics, including internal donors and solvents, are routinely quantified by sample dissolution and gas chromatographic analysis. NMR spectroscopy is an alternative technique. The surface structure of organic components can be readily studied by FTIR spectroscopy [19]. Particle size distribution is measured by laser diffraction of an oil slurry of the catalyst. Polymerization testing and polymer characterization are described in Chapters 10 and 2, respectively.

9.11   Questions

1. Which catalyst, a 4th generation supported catalyst or a precipitated $TiCl_3$ catalyst is more sensitive to oxygen, and why?
2. A researcher made a fourth generation catalyst preparation in the equipment in Figure 9.3, except that the equipment was simplified by using a magnetic stir bar in place of an agitator shaft. When the catalyst was isolated and examined under the microscope, irregular particle sizes and dust were observed. What might be the reason for this?

3. Using the catalyst example in section 9.8.2, and the nominal prices below, calculate the raw materials cost for producing one pound of catalyst. Assume that the final weight of the catalyst is equivalent to the weight of magnesium ethoxide charged to the flask. Next calculate the total weight of reagents used. What can one say concerning the eco-consequences of the process design?

Magnesium ethoxide	$5.00/lb
Titanium tetrachloride	$1.00/liter
Toluene	$1.00/liter
Heptane	$1.00/liter
Di-*n*-butylphthalate	$2.00/lb

4. A catalyst is washed five times with hexane. At each washing the concentration of chlorinated compounds in the decanted hexane decreases by a factor of four. Can you think of a washing strategy to reduce the amount of spent hexane that needs to be recovered by distillation?
5. After completing a preparation of a supported catalyst, the researcher notices the formation of solids in the waste bottle, which contains all the effluents from the synthesis. What are the implications for manufacturing process design?
6. Why is an oil heating bath preferred over an electric heating mantle?
7. Why is an oil bath with dry ice chips preferred over a water ice bath?

References

1. Schlenk glassware and Schlenk operations refer to specialized glassware for air-free reactions and manipulations developed by Wilhelm Schlenk, a German chemist, in the early 20th century.
2. Glassware suppliers: www.aceglass.c om; www.kimble-chase.com
3. Equipment suppliers: www.coleparmer.com; www.fishersci.com
4. For example Ace catalog number 6955–08.
5. A dropping funnel with a PFTE needle valve stopcock for precise adjustments is preferred; see Ace catalog number 7298–05, for example.
6. For example, Ace sparger tubes catalog number 6452 types.
7. For example, Ace threaded adapters catalog number 5030 types.
8. The use of mercury thermometers is discouraged in the modern laboratory.

9. Type B is for flammable solvents and type D is for flammable metals, i.e. aluminum alkyls.
10. An acid vapor suppression extinguisher is to suppress HCl fumes from TiCl4 spills.
11. J.A. Young, *J. Chem. Educ.* vol 84, pg. 1760, 2007.
12. The nonpyrophoric limit of DEAC in hexane or heptanes is 13 wt%. See: "Pyrophoricity of Metal Alkyls", AkzoNobel Polymer Chemicals, OMS 03.322.04 August 2008: http://www.akzonobel.com/hpmo/system/images/AkzoNobel_Pyrophoricity_of_Metal_Alkyls_ma_glo_eng_tb_tcm36–16299.pdf; see also D. Malpass, *Handbook of Transition Metal Polymerization Catalysts* (R Hoff and R Mathers, editors), John Wiley & Sons,) 551 2010.
13. See, for example, http://www.envirosafetyproducts.com/product/Stanco-Nomex-Lab-Coat.html
14. T. Yano, M. Hosaka, M. Sato, K. Kimura, US patent application 2010/0190938A1; M. Hosaka, T. Yano, H. Kono, US patent 7704910B2; M. Hosaka US Patent No. 7005399B2.
15. G. Morini, E. Albizatti, G. Balbontin, G. Baruzzi, A. Cristofori, US Patent No. 5723400; E. Albizatti, P.C. Barbe, L. Noristi, R. Scordamaglia, L. Barino, U. Gianinni, G. Morini, US Patent No. 4971937.
16. G. Morini, G. Balbontin, Y.V. Gulevich, H.P.B. Duijghuisen, R.T. Kelder, P.A.A. Klusener, F.M. Korndorffer, US patent 6818583 B1; G. Morini, G. Balbontin, G. Vitale, US Patent No. 7202314 B2.
17. P. Hermans, P. Henrioulle, US Patent No. 4210738.
18. C.F. Petry, L.B. Capeletti, F.C. Stedile, J.H.Z. dos Santos, D. Pozebon, *Analytical Sciences*, vol. 22, pp. 855–859, 2006.
19. U. Makwana, D.G. Naik, G.Singh, V. Patel, H.R. Patil, V.K. Gupta, *Catalysis Letters*, vol. 131, pp. 624–631, 2009.

10

Polymerization Catalyst Testing

10.1 Introduction

This chapter describes a methodology for propylene polymerization testing of Ziegler-Natta catalysts. It is limited to polypropylene homopolymer; copolymer testing can be considerably more complex. The *raison d'être* of the catalyst is its polymerization performance. For both quality control purposes and for laboratory investigations polymerization performance is measured by a batch method. This requires a scale convenient to the laboratory, typically in a reactor size of a few liters. Literature descriptions of catalyst polymerization testing usually provide only a brief description of the equipment and procedures [1]. Early descriptions employed glass round bottom flasks, mercury manometers, and magnetic stirring bars for low pressure tests and rocking autoclaves for high pressure testing [2]. These types of equipment are, generally speaking, no longer used, having been replaced by modern autoclaves.

Catalyst performance may be evaluated by two principal outcomes: catalyst activity and polymer properties. However, these must be specified under a particular set of conditions. There are a variety of ASTM methods for testing polypropylene properties,

217

and these are discussed in chapter 2 on polymer characterization. However, there is no standard test for catalyst performance. This may be due to historical development of a variety of industrially important catalysts (see Chapter 4) and to the variety of industrial processes to produce polypropylene (Chapter 8). Each catalyst type required its own test conditions, both to mimic the industrial conditions for which it was developed, and to optimize its own performance. So, for example, the external donor ideally suited for one catalyst, may not be optimal for another catalyst, or the polymerization time for one catalyst may not reflect the industrial residence time of another catalyst. For purposes of this section, polymerization test conditions have been selected that are reasonably suitable to the catalysts whose syntheses are described in Chapter 9.

Almost all of today's industrial plants produce polypropylene in neat propylene (bulk polymerization) or in gas phase propylene (or using a dual process in both phases). That is, no liquid diluent is used. Accordingly, many laboratories employ these conditions for testing catalysts. However, testing in a research or quality control laboratory may also be done in an inert hydrocarbon diluent, such as hexane, in order to have milder test conditions that are more attractive to the laboratory. This is called slurry testing [3], and it originated with the early polypropylene processes which were operated in a diluent to selectively dissolve the relatively large amount of atactic polymer produced by first generation catalysts (Chapter 4).

Working with a diluent provides some additional advantages in the laboratory compared to testing in liquid propylene [4]. First, the pressure of the polymerization is reduced; testing with neat propylene requires a pressure of ~40 atmospheres to liquefy the propylene at 70 °C versus about 7 atmospheres for a reasonable concentration of gaseous propylene dissolved in a hexane diluent at the same temperature. A reference curve of propylene vapor pressure vs. temperature is shown in Figure 10.1. Second, the working inventory of propylene in the reactor is greatly reduced, which is a safety benefit for the laboratory. Third, reducing the propylene concentration mitigates the need for separately prepolymerizing the catalyst in order to maintain polymer morphology, especially for highly active catalysts. This simplifies the polymerization procedure. A final advantage is that, with an in-line mass flow sensor, it is much easier to study the kinetics of a slurry polymerization than to study the kinetics of polymerization in liquid propylene. Accordingly, the slurry polymerization procedure has been chosen for this section.

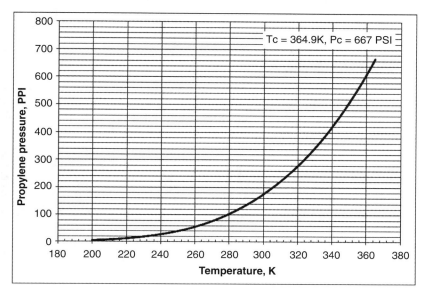

Figure 10.1 Propylene vapor pressure vs.temperature [5]. data calculated from the *Chemical Properties Handbook*, C.L. Yaws editor, McGraw Hill, 1999, according to the following formula where P = mm Hg, and T = K, and then P is converted to PSI:

$$P = 10^{\wedge}(24.5390 + (-1.5072E+03/T) + (-6.4800E+00*(log(T))) + (-4.2845E-11*T) + (5.4982E-06*(T^{\wedge}2)))$$

10.2   Facility Requirements

Setting up a polymerization testing facility requires a substantial commitment in planning, time, and money. A schematic diagram of a polymerization testing apparatus is shown in Figure 10.2. While the diagram is intended to be informative, it is not meant as a detailed equipment design. The polymerization test requires a stainless steel autoclave and associated feed lines for the monomer, hexane, hydrogen, and nitrogen. It also requires equipment for preparing the catalyst charge. Finally, the equipment must be operated in a safe working environment, requiring proper ventilation, alarms, and emergency response. Detailed laboratory design is beyond the scope of this book, and interested readers should consult further with equipment suppliers to develop a suitable laboratory facility [6].

An important auxiliary item for manipulating catalyst samples and for preparation of the catalyst charge is the inert atmosphere glove box. This is a highly recommended tool, but adds significant

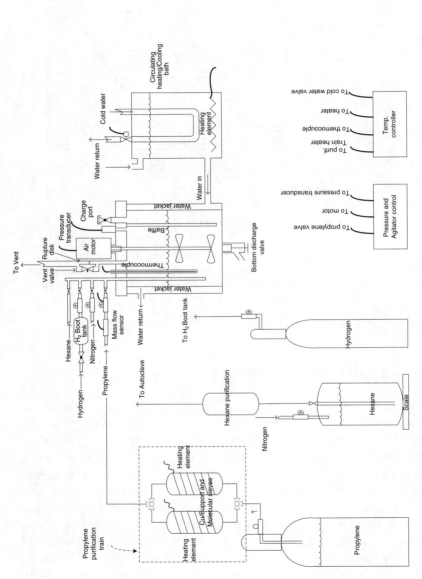

Figure10.2 Polymerization test apparatus schematic diagram. Basic components of the apparatus include propylene storage, purification and delivery system, diluent storage and delivery, hydrogen delivery, autoclave and controller, and heating/cooling, with controller. Components are not drawn to scale.

Figure 10.3 Vacuum atmospheres inert atmosphere Glove Box. The inert atmosphere glove box provides a convenient means of manipulating highly air sensitive materials. Materials are admitted to the glove box via the double door port on the right side, which is completely evacuated and then refilled with an inert gas. Photo Courtesy of W.E. Summers III.

cost. It permits precise catalyst handling under stringent conditions. A representative example is shown Figure 10.3. Having a glove box facilitates accurate weighing to 0.1 mg, permits safe preparation of the catalyst/cocatalyst/external done suspension, and greatly reduces the chance of degrading the catalyst components from contact with moisture or oxygen.

10.3 The Autoclave

The main test apparatus is, of course, the autoclave. A typical reactor is shown in Figure 10.4. It should be designed in cooperation with the vendor. Stainless steel (type 316) construction is recommended. The interior surfaces should be smooth and polished. An external heating/cooling jacket is recommended over an internal coil to reduce places where catalyst particles can lodge. The agitator shaft should be magnetically coupled to the motor. This provides a seal free from mechanical couplings and the possibility of seal oil contamination into the reactor. An air-driven motor is preferred for

Figure 10.4 One gallon polymerization test reactor. The steel autoclave is supported by a hydraulic system which lowers the reactor body while the head remains stationary. Normally the reactor is emptied by opening the bottom discharge valve. Photo Courtesy of W.E. Summers III.

safety purposes. A pitched blade impeller design enables the agitator to create a downward flow, pushing the slurry down and outward during mixing. See Figure 10.5. In combination with a robust magnetically coupled drive unit, a good vortexing action can be achieved which draws gases into the slurry. Polymerization occurs only at the surface of the catalyst, so the local concentration of monomer and hydrogen are critical. Some vendors offer a hollow agitator shaft that draws gases from the headspace to the bottom of the vessel by a venturi effect. The reactor design should include a

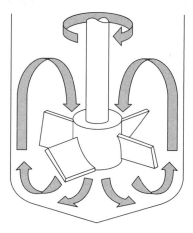

Figure 10.5 Vortex from pitched blade turbine impeller [7]. Reproduced by permission of Autoclave Engineers, a division of Snap-Tite, Inc., E Catalog pg 6: http://www.autoclaveengineers.com/ecat/ecat_update.htmL.

baffle for additional mixing. In some cases the dip tube or thermocouple can serve this function.

Solids addition and discharge are key design components. The reactor charge port must be wide enough to conveniently allow the insertion of the syringe or flexible tubing that will be used to charge the catalyst suspension. The inclusion of a bottom discharge valve allows the operator to conveniently discharge the polymer slurry and to keep the autoclave closed between uses. The design of the discharge valve must take into account the size of the polymer particles that it will need to handle. Its material of construction must be compatible with all the reagents. A piston valve is favorable for achieving a flush interior bottom surface profile.

The support for the reactor assembly is important for operational efficiency. The assembly should be supported at the reactor head (see Figure 10.4). When it is necessary to open the reactor, the body is lowered while the head remains in place, eliminating the need to remove head fittings. A hydraulic jack is recommended for lowering the body, although it can be done manually for small reactors up to 2 liters. The height of the reactor assembly should allow sufficient space for a receiver below the bottom discharge valve, and the valve handle should be high enough from the floor to allow convenient operation.

The dimensions of the ventilation hood need to be sufficient for easy access to the autoclave. This includes the space needed for convenient charging and discharging. The propylene tank and purification assembly need to be located in a well-ventilated area. An area adjacent to the hood housing the polymerization reactor is preferred.

10.4   Key Equipment Items

- Ventilation hood
- 2-L stainless steel jacketed autoclave with bottom discharge valve
- Circulating heating/cooling fluid supply

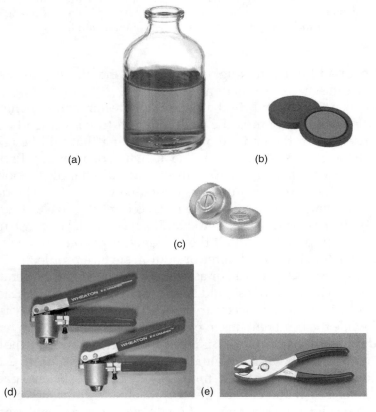

Figure 10.6 Septum Bottle and Accessories (a)Wheaton Septum Bottle, (b) Teflon Coated Septa, (c) Caps, (d) Crimper, and (e) Decrimper Reproduced by permission of Wheaton Industries, Millville, NJ USA.

- Temperature controller
- Propylene purification and delivery system
- Hydrogen delivery system
- Diluent (hexane) purification and feed system
- Nitrogen delivery system
- Septum bottles, Teflon lined septa, crimping tool (Figure 10.6)
- Inert atmosphere glove box
- Precision balance +/− 0.0001 g.
- Gas-tight syringe with locking valve and needle [8]
- Grounded metal discharge pail
- Large Buchner funnel and filter flask

10.5   Raw Materials

- *Propylene:* Polymerization grade propylene is required. Typical specifications are shown in Table 10.1. The propylene feed is further purified by columns as shown in Figure 10.2. The purification columns commonly employ activated copper on an alumina

Table 10.1 Polymerization grade propylene allowable impurities levels [9].

Component	Maximum (ppm)
Propane, butane	1000
Ethylene	100
Water	5
Oxygen	5
Sulfur	1
Carbon dioxide	5
Carbon monoxide	0.3
Acetylene	3
Carbon oxysulfide	0.1
Arsine	0.1

support and molecular sieves [10]. The former is purchased as copper oxide on alumina and reduced to activated copper by heating the column in flowing hydrogen. The copper removes carbon monoxide, alcohols, amines and sulfur containing impurities, while the molecular sieves remove water and methanol. Special adsorbents can be purchased for arsine and phosphine removal, which are especially potent catalyst poisons [11]. It is interesting to note that the catalyst poisoning properties of carbon monoxide are used industrially to rapidly effect emergency shutdown in commercial polymerization reactors (see section 8.9). The copper column can be regenerated in place using heat and flowing hydrogen. Similarly, the molecular sieves can be regenerated using heat and flowing nitrogen. To avoid interruptions two columns can be set up in parallel, so that if one column is being regenerated, the other is available for use. This is shown in Figure 10.2.

- *Hydrogen:* High purity hydrogen can be used without further purification.
- *Nitrogen:* High purity nitrogen, or boil off from liquid nitrogen can usually be used without further purification.
- *Hexanes:* Hexanes (various isomers) are preferred over n-hexane due to peripheral neuropathy concerns [12]. It should be dried and purified by passage through a column of 3A or 4A molecular sieves and silica gel.
- *Triethylaluminum (TEAL) and Diethylaluminum Chloride (DEAC):* These materials are used as purchased. It is more convenient to obtain solutions in hexane at about 10 wt%, which is below the NPL (non-pyrophoric limit). Safety aspects on handling are discussed in Chapter 5. Detailed handling instructions come with the sample bottles.
- *External Donor:* The selection of the external donor is dependent upon the catalyst. For most modern supported catalysts this will be an organosilane. The silane is used as received. If desired, a stock solution in hexane can be prepared in a septum bottle and

stored in the glove box. The example below uses chlorotriethoxysilane, based on the reference for the supported catalyst synthesis in Chapter 9 [13].

10.6   Polymerization Conditions

As noted above, test conditions reflect anticipated use in a commercial polypropylene plant, as much as practical, for a laboratory test. The most profound difference is that commercial processes are continuous in all aspects: catalyst, cocatalyst, external donor and propylene are fed continuously and polymer is removed continuously. By contrast, the laboratory test is a batch (liquid propylene) or semi-batch process (continuous feed of propylene gas). Charge ratios, cocatalyst selection, temperature and time are chosen accordingly. Hydrogen charge is chosen so as to produce a polymer in the melt flow range of interest. In a commercial plant the catalyst is often prepolymerized before entering the main process vessel. A separate prepolymerization is not essential under the more dilute conditions of the hexane slurry test. The catalyst is introduced into the reactor at low temperature and propylene concentration, so, as the temperature and pressure are increased there is an *in-situ* prepolymerization that occurs during the minutes required for the system reaches final operating conditions. Basic conditions for the two catalysts described in Chapter 9 are shown in Table 10.2.

Table 10.2 Polymerization test conditions.

Item	Precipitated TiCl$_3$ Catalyst	MgCl$_2$ Supported Catalyst
Catalyst Charge, mg	35	20
TEAL, mmol	0	2.0
DEAC, mmol	2.0	0
Chlorotriethoxysilane, mmol	0	0.20
Hydrogen, STP mL	500	500
Time, h	3.0	1.0
Temperature, C	80	70

10.7   Autoclave Preparation

It is critical to begin with a polymerization reactor that is clean and in good condition. Residues, scratches and irregular surfaces can trap catalyst and growing polymer particles and distort the polymerization test results [14].

1. Pressurize the reactor to ~ 150 psig with nitrogen and pressure test the reactor for leaks. There should no pressure drop over 30 minutes.
2. Flush the propylene gas line briefly into the reactor.
3. Vent the reactor and purge with nitrogen through the charge port for at least 15 minutes. Stop the nitrogen and close the vent.
4. Add one liter of hexane from the storage tank to the autoclave by a transfer line and begin agitation.
5. Degas the hexane charge by pressurizing with nitrogen to 50 psig and depressurizing four times.
6. Set the reactor temperature to 40 °C.
7. Add 20% of the cocatalyst charge in Table 10.2 *via* syringe to the reactor while flowing nitrogen from the addition port, then close the port [15].
8. Pressurize the hydrogen boot cylinder with ~ 200 psig hydrogen.

10.8   Polymerization Test Procedure

1. In a glove box equipped with a precision balance, weigh out the catalyst per Table 10.2, into a tared 50-mL glass septum bottle. Carefully record the net weight to 0.1mg [16]. Add 20 mL hexane, then the remaining cocatalyst, and then the external donor per Table 10.2. Tightly close the bottle with a Teflon lined septum and crimp cap or septum screw cap and shake to mix the components. Prepare a second bottle containing 30 mL hexane and cap it as well.
2. Remove both bottles from the glove box.
3. Purge a 50-mL gas tight syringe with a valve and 6-inch needle several times with nitrogen. Inject the septum bottle with ~ 20 mL nitrogen from the syringe, shake

the bottle and withdraw the slurry into the syringe while shaking. Close the syringe valve.

4. Transfer the syringe to the autoclave and open the charge port under a gentle nitrogen flow. Open the syringe valve, and inject the slurry into the reactor while maintaining the nitrogen flow out of the charge port. Using the septum capped bottle of hexane, immediately draw about 20 mL into the syringe, inject it into the catalyst bottle, shake and draw back into the syringe. Add this rinse solution into the reactor. Inspect the bottle to be sure the catalyst transfer is complete. Repeat the hexane wash if necessary.

5. Close the charge port and close the nitrogen inlet.

6. Start the agitation and increase it to 600 rpm.

7. Add the hydrogen from the boot tank per Table 10.2. For a 200-mL boot tank, 500 STP mL H_2 corresponds to a pressure drop of ~ 40 psig. Set the reactor temperature for 70 °C and gradually pressurize the autoclave to 120 psig with propylene.

8. Maintain the pressure at 120 psig per Table 10.2 at 70 °C. After the prescribed time, begin cooling and vent the reactor *very* slowly (rapid venting can cause slurry to enter the vent system). Continue venting until only a few psig pressure remains, cool the contents to 25 °C and discharge with slow agitation using the bottom valve, receiving the slurry in a grounded metal bucket, using the remaining pressure to motivate flow of the slurry [17].

9. Rinse the reactor with a liter of hexane, agitating for five minutes, then discharge and combine with the product slurry.

10. Collect the polymer by filtration in a well-ventilated hood. A large Buchner funnel and filter flask are suitable. Transfer the filter cake into a tared metal pan and air dry in a hood or in a forced air oven. Do not use an electrically heated oven unless it is rated to handle flammable vapors. When the sample pan comes to constant weight this is the Insoluble Polymer yield.

11. Weigh the total filtrate, take a measured aliquot (5–10% of the filtrate), and evaporate it to dryness in a hood in a tared disposable pan on a steam plate.

12. Calculate the soluble polymer yield:
    Soluble polymer = g in pan/(fraction of filtrate evaporated)
13. Calculate the Total Hexane Solubles:
    Total Hexane Solubles (%) = 100 x g Soluble Polymer/g Total Polymer
    Where g Total Polymer = g Soluble Polymer + g Insoluble Polymer
14. Calculate catalyst Activity:
    Activity = (g Insoluble Polymer + g soluble polymer)/g Catalyst

Procedural Notes

1. The catalyst charge is an approximate guideline. It may need to be adjusted depending upon the particular catalyst activity. Adjust the cocatalyst charges accordingly.
2. The polymerization reaction is exothermic, so there may be temperature overshoot observed. This can be compensated by initially setting the temperature controller at a lower temperature. Programmable controllers often have a learning mode that allows them to modify the control protocol to account for exotherms.
3. If the propylene line is equipped with a mass flow sensor, then the kinetics of the reaction can be monitored by recording the flow rate vs. time [18].
4. It is important for the reactor to be thoroughly agitated to aid in propylene and hydrogen diffusion into the hexane. This requires proper impeller design, correct height placement, and speed to develop the vortexing action needed. Two sets of impellers are recommended.
5. Increasing the hydrogen charge tends to increase the polymerization activity if the polymer melt flow index is ~ 2 g/10 min or less.

10.9 Reproducibility

Testing of polymerization catalysts may have many purposes. It may be connected with an exploratory program to develop new types of catalysts. It may be for developing incremental improvements in existing catalysts. It may be for quality control or comparative purposes in the manufacture of commercial catalysts. Depending upon the purpose, demands on accuracy and reproducibility may differ. In all cases it is presumed that the testing program is of long duration, *i.e.* that results will be compared between tests made over months or years. This requires long term consistency and reproducibility of the test method.

Consistent results depend, first of all, upon consistent raw materials. Hence it is important to use reliable raw materials and to purify them to the levels required by the particular catalysts being tested. Secondly, consistency depends upon a clearly defined test method. Carefully constructed procedures leave little room for operator variability. The test methods include raw materials purification, calibration of equipment, and training of the operator.

It is a good practice to have a control test for the polymerization system. The control test is made on a periodic basis to judge the status of the test method. A control catalyst is established for which there is a sufficient inventory for many tests, and it is carefully stored to preserve its integrity. Polymerization results are plotted on a Shewhart chart and control limits are set based on statistical principles [19]. When results are outside of the control range an investigation must be made into the cause(s) and those causes corrected prior to testing new experimental samples. For example, low activity in a control test might be due high water content in hexane, which, in turn, may be due to exhaustion of the purification column for the hexane. Alternatively, variation might be due to an analytical balance calibration problem. Without statistical control methods in place it will be very difficult to compare results obtained months, or even weeks, apart.

Reproducibility of the test improves with attention to detail, especially accurate catalyst weight, complete delivery of the catalyst charge, and the absence of contaminants. An achievable goal for activity is a standard deviation of +/− 5%.

Physical property testing of the polymer is discussed in Chapter 2.

10.10   Testing Metallocene Catalysts

The polymerization testing of metallocene catalysts is fundamentally the same as for ZN catalysts. Methylaluminoxane (MAO) is typically substituted for TEAL, and this introduces a small amount of toluene into the system. Alternatively, MAO may be incorporated into the solid catalyst synthesis, as described in section 6.5. In such cases the solid catalyst may be pyrophoric. If all the MAO is included in the solid catalyst, then a small quantity of TEAL or TIBAL should be added into the polymerization reactor as a poison scavenger, prior to the introduction of the solid catalyst. Some metallocene catalysts substitute MAO with cation-forming cocatalysts such as perfluorinated boranes (section 6.4.2). External donors are usually omitted from the polymerization, and molecular weight can be very sensitive to added hydrogen. Purification of raw materials is at least as important for metallocenes as for ZN catalysts because every metal center is designed to be an active site.

10.11   Questions

1. A supported catalyst containing 2.5 wt% Ti produces 40,000 kg polypropylene/kg catalyst in commercial use. If 100% of the titanium is catalytically active, how many ppm of carbon monoxide (CO) added to the propylene feed would be sufficient to kill half the catalyst, assuming each molecule of CO poisons one titanium center? If only 10% of the total titanium is catalytically active, how many ppm of CO would be sufficient, assuming CO reacts only with active centers? How do these values compare with the value in Table 10.1?

2. A precipitated titanium trichloride catalyst containing 25% Ti produces 5,000 kg polypropylene/kg catalyst in commercial use. If 100% of the titanium is catalytically active, then how many ppm of carbon monoxide (CO) added to the propylene feed would be sufficient to kill half the catalyst, assuming each molecule of CO poisons one titanium center? Compare the value with Question 1. What can one say about the relative sensitivity to propylene impurities of supported catalysts and titanium trichloride catalysts?

3. How can one efficiently test the effect of changes in polymerization temperature on reaction kinetics in a slurry test?

4. Considering the cocatalyst aluminum alkyl, what might be one reason that alcohols are not strong catalyst poisons like CO?

5. In a slurry propylene polymerization using the protocol of this chapter, 20 mg of supported catalyst gave a yield of 10,000 g polymer/g catalyst. Assuming the number average polymer molecular weight (M_n) is 100,000, and that each chain is terminated by a hydrogen atom at each end, how many STP mL H_2 are consumed during the polymerization? How does this compare with the total STP mL H_2 charged?

6. What would be the effects of flushing the autoclave with propylene, instead of nitrogen, during the catalyst charging step?

7. Suggest a method of cleaning the interior of an autoclave which has accumulated a film of polypropylene.

8. Why is 20% of the cocatalyst charged to the hexane in the reactor in step 7 of the polymerization procedure?

9. Why are external donors omitted from the polymerization recipe when using a metallocene catalyst?

References

1. R. Lieberman, C. Stewart, *Encyclopedia of Polymer Technology*, J. Wiley and Sons, vol. 11, pp. 312–313.

2. J. Boor Jr., *Ziegler-Natta Catalysts and Polymerizations*, Academic Press, 1979, pp. 61–63.

3. The term "slurry" testing is a bit of a misnomer. It refers to polymerization in an inert hydrocarbon diluent. However, polymerization either in inert hydrocarbon or in propylene monomer results in a slurry of polymer *in-situ*. Upon discharge from the reactor the term has more meaning; in an inert hydrocarbon a slurry is discharged, whereas with propylene monomer the unreacted propylene is vented off prior to discharge of a powder product.

4. It is also possible to test catalysts under batch gas phase conditions in a similar pressure range as slurry testing. However, this requires a seed bed of polymer or other inert material onto which to deposit the catalyst and careful dispersion of the catalyst/cocatalyst charge onto the seed bed. A special helical impeller is required to keep the polymer bed agitated, and gas phase polymerization is more prone toward developing hot spots due to lower heat transfer characteristics.

5. *Chemical Properties Handbook*, C.L. Yaws editor, McGraw Hill, 1999.

6. See for example, www.autoclaveengineers.com ; www.pdcmachines.com; www.parrinst.com ; http://pressurereactor.buchiglas.ch/home.htmL

7. http://www.autoclaveengineers.com/ecat/ecat_update.htmL

8. For example, see Beckton Dickinson for Luer-Lok™ devices.

9. For a more complete, and more stringent set of specifications see *Polypropylene Handbook*, 2nd edition, Nello Pasquini editor, Hanser Publishers 2005, pg. 376.

10. D.M. Haskell, B.L. Munro, US patent 3676516; For a commercial supplier see http://www.psbindustries.com/pdf/Selexsorb%20CD%20 Data%20Sheet.pdf

11. Selexsorb AS supplier: http://www.coastalchem.com/PDFs/ Engelhard/SelexsorbAS.pdf

12. http://www.cdph.ca.gov/programs/hesis/Documents/nhexane_ med_guide.pdf

13. T. Yano, M. Hosaka, M. Sato, K. Kimura, US patent application 2010/0190938A1; M. Hosaka, T. Yano, H. Kono, US patent 7704910B2; M. Hosaka US Patent No. 7005399B2.

14. Use plastic or brass tools for cleaning surfaces, to avoid scratching the steel.

15. The purpose of this step is to scavenge impurities from the hexane prior to exposure to the catalyst.

16. If a glove box is not available, then in a glove bag prepare a 1 wt% suspension of catalyst in mineral oil in a septum bottle. Shake the bottle well and withdraw the appropriate amount by syringe.

17. The use of a grounded metal bucket is important to prevent the possibility of a static discharge from igniting the hexane vapors as the reactor is emptied.

18. Mass flow sensors are available from: www.brooksinstrument.com; www.sierrainstruments.com

19. D.C. Montgomery, *Introduction to Statistical Quality Control*, Wiley, 2008; M.J. Chandra, Statistical Quality Control, CRC Press LLC, 2001.

11

Downstream Aspects of Polypropylene

11.1 Introduction

Polypropylene's circuitous journey from the manufacturer to the consumer typically begins with the raw polymer being formulated with an antioxidant, melted, pelletized (see Figure 1.3), packaged and transported to processors, usually by rail car. Processors introduce other additives (specific to the end-use), melt the polymer again and shape the molten polymer into myriad practical articles employing a variety of fabrication techniques (see section 11.3). Polypropylene reaches the consumer in an enormous range of goods. It could take the form of a casing for a power tool in the garage, or fibers in the living room carpet, or the agitator in the family washing machine, or a microwavable container in the kitchen, or hundreds of other items that we encounter everyday in our homes, automobiles and workplaces. In many instances, the consumer may be completely unaware of the impact that polypropylene has on daily life, but would certainly notice if these items were suddenly unavailable. When the article's useful life is over, the question of how to dispose of the polymer must be addressed.

During several of the processing steps mentioned above, polypropylene is exposed to harsh conditions that can compromise its molecular integrity. The polymer is especially vulnerable when it is heated to temperatures of >200°C and subjected to stress/strain forces that can cause bond scission and other chemistries that degrade the polymer.

However, when stabilized with an antioxidant, polypropylene is a durable, tough, chemically resistant material and is not readily biodegradable. Though biodegradable alternatives to polyolefins have become commercially available of late, the practicality of such products under present-day market conditions remains tenuous. Unless heavily subsidized or mandated by government, biodegradable alternatives to polypropylene and other plastics will need to become cost-competitive to compete in the free market [1] and that does not appear likely in the foreseeable future.

Additives, fabrication methods, biodegradable alternatives and environmental issues are far too broad and complex for detailed discussions here. We will, however, provide perspectives on downstream aspects, addressing such essential questions as:

- Why are additives necessary?
- Why are antioxidants so crucial for polypropylene?
- What are the principal fabrication methods used for polypropylene?
- What are the "green" alternatives to polypropylene derived from petroleum?
- What happens to an article made from polypropylene at the end of its life cycle?

References for detailed information about the topics discussed in this chapter are contained in each section.

11.2 Additives

Additives are vital to the performance of polypropylene. Indeed, modern polymers in general would not be viable without attributes that additives impart to the fully formulated resin.

Typically, after virgin polypropylene exits industrial reactors, it is melted and an antioxidant added. The molten, stabilized polymer is extruded, cooled and cut into pellets. Processors introduce other additives (see partial list below). Finally, the pellets are once again melted and the product is fabricated *via* a variety of techniques into innumerable articles that are purchased by the consumer.

The repeated melting and resolidification of the product is termed its "heat history." As a consequence of these different processing steps, the polymer undergoes degradation which can adversely affect its mechanical properties. For example, melt flow rate increases (*i.e.*, molecular weight decreases). Table 11.1 shows the effect of multiple extrusions on melt flow rate of unstabilized polypropylene homopolymer. These results demonstrate why it is essential to add an antioxidant as soon as practical to prevent oxidative degradation. Broadly speaking, additives are introduced:

- to stabilize the polymer,
- to make the polymer easier to process, and/or
- to enhance its end use properties.

To achieve these objectives, many types of additives are required. A *partial* list is given below:

- *antioxidants*
- *antistatic agents*
- *nucleating agents*
- *light (UV) stabilizers*
- *lubricants*
- *fillers*
- *antimicrobials*
- *slip agents*
- *reinforcing agents*
- *acid scavengers*
- *flame retardants*
- *anti-blocking agents*
- *colorants*

Discussions of the many additives used in polypropylene are clearly outside the scope of an introductory text. However, because of the critical importance of preventing oxidative degradation of polypropylene, we will survey principal aspects of antioxidants. There are many sources for specifics on additives to which the reader is referred for in-depth discussions [2–9]. See especially the thorough handbooks edited by Zweifel [2, 3] and a text by Fink [4].

Table 11.1 MFR of homopolymer vs. multiple extrusions.

Number of Extrusions*	MFR** (dg/min)
1	7
3	25
5	>30

* At ~270°C.
**Melt flow rate under ASTM D1238-04c, Condition 230/2.16.

Why is polypropylene so susceptible to autoxidation? As shown in Chapter 1, each repeating unit in polypropylene contains a primary (1°), a secondary (2°) and a tertiary (3°) carbon atom. Bond disassociation energies (D) for C-H bonds decrease in the series: 1° > 2° > 3°, as illustrated below (other H atoms not shown):

$$\sim\text{C-C-H} \quad \xrightarrow[\text{or catalysts}]{h\nu \ (light)} \quad \sim\text{C-C}\bullet \quad + \quad \text{H}\bullet \quad D \sim 420 \text{ kJ/mole} \qquad (11.1)$$

$$\sim\overset{\displaystyle |}{\underset{\displaystyle C}{\text{C}}}\text{-C-H} \quad \xrightarrow[\text{or catalysts}]{h\nu \ (light)} \quad \sim\overset{\displaystyle |}{\underset{\displaystyle C}{\text{C}}}\text{-C}\bullet \quad + \quad \text{H}\bullet \quad D \sim 401 \text{ kJ/mole} \qquad (11.2)$$

$$\sim\overset{\displaystyle C}{\underset{\displaystyle C}{|\ \ |}}\text{C-C-H} \quad \xrightarrow[\text{or catalysts}]{h\nu \ (light)} \quad \sim\text{C-C}\bullet \quad + \quad \text{H}\bullet \quad D \sim 390 \text{ kJ/mole} \qquad (11.3)$$

Hence, the abundant tertiary C-H bonds in polypropylene undergo homolytic scission more readily than secondary or primary C-H bonds. For this reason, polypropylene is vulnerable to free radical formation at the tertiary carbon atom, which can subsequently interact with atmospheric oxygen to form peroxy radicals. These peroxy radicals may abstract hydrogen, *e.g.*, from ambient water or from the polymer backbone, to produce unstable hydroperoxides. In turn, hydroperoxides decompose to produce polymeric fragments that contain a variety of oxygenated organic functional groups (alcohols, ketones, aldehydes, etc.). These free radical reactions are rather complex and uncontrolled and result in a loss of mechanical strength due to the decrease in molecular weight. Exemplary reactions that unstabilized polypropylene macroradicals may undergo

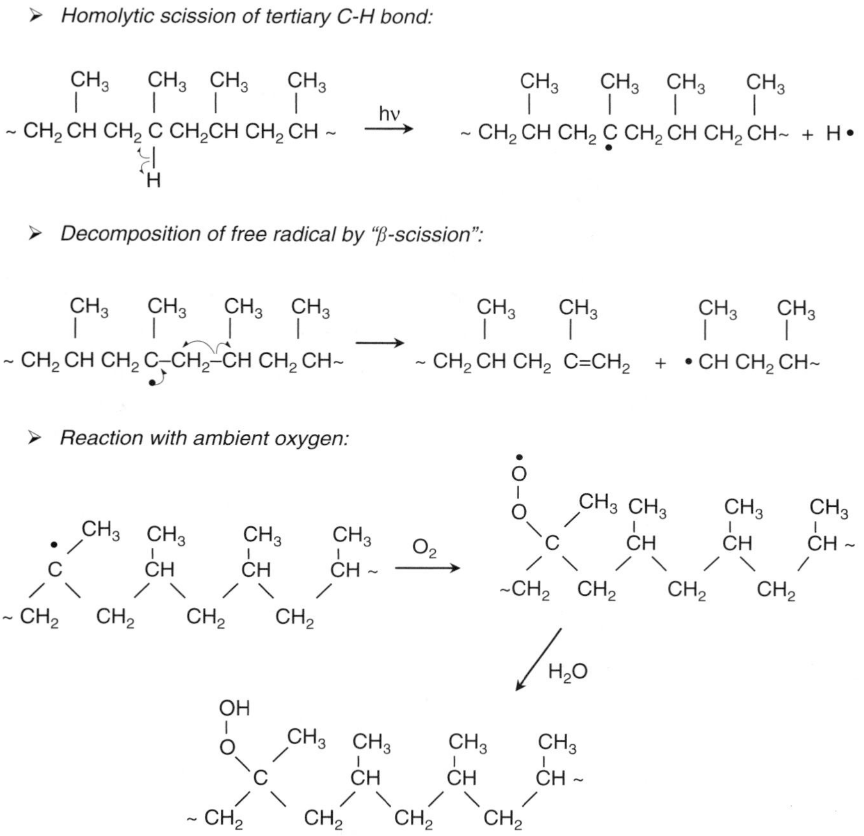

➢ *Homolytic scission of tertiary C-H bond:*

➢ *Decomposition of free radical by "β-scission":*

➢ *Reaction with ambient oxygen:*

Figure 11.1 Exemplary free radical degradation reactions of polypropylene. Macroradical containing a hydroperoxide group decomposes to polymeric fragments containing organic functional groups (alcohols, ketones, aldehydes, etc.).

are illustrated in Figure 11.1 (These are but a few of the many free radical reactions by which polypropylene may degrade. See references 2–4, 10 and 11 for more information.)

Because of reactions illustrated in Figure 11.1 and discussed above, antioxidants are crucial to downstream applications of polypropylene. Antioxidants interrupt the cycle of degradation by scavenging or decomposing free radicals. Among the most effective antioxidants are hindered phenols, which account for about 56% of the total market for antioxidants sold into the global plastics industry as shown in Figure 11.2. Phosphite esters are second in importance

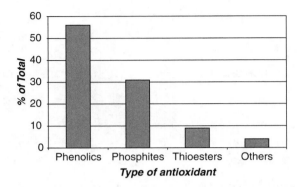

Figure 11.2 Global consumption of antioxidants in plastics. Source: *Plastics Additives Handbook*, H. Zweifel, R. Maier and M. Schiller, editors, Hanser Gardner Publications, Inc., Cincinnati, 6th Ed., p. 2, 2009.

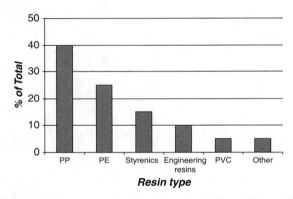

Figure 11.3 Global consumption of antioxidants by resin. Source: *Plastics Additives Handbook*, H. Zweifel, R. Maier and M. Schiller, editors, Hanser Gardner Publications, Inc., Cincinnati, 6th Ed., p. 3, 2009.

as antioxidants. Because they act synergistically, hindered phenols are often used in conjunction with phosphites. Together, hindered phenols and phosphites account for more than 85% of antioxidants sold into the plastics industry as shown in Figure 11.2. The polyolefins industry is the largest consumer of antioxidants, accounting for about ⅔ of the total global consumption as shown in Figure 11.3. More specifically, polypropylene is the single largest consumer of antioxidants, accounting for about 40% of the total market.

Antioxidants are classified as primary or secondary, depending upon how they react. Hindered phenols are primary antioxidants and function by donating a hydrogen to convert a peroxy radical to a hydroperoxide. Phosphites are among antioxidants that are termed "secondary" and function as hydroperoxide decomposers. The ultimate outcome of these reactions is to convert macroradicals to derivatives that are less destructive to the polymer.

A hindered phenol commonly used as an antioxidant is 2,6-di-*tert*-butyl-4-methylphenol (also known as butylated hydroxy toluene or "BHT"). Many hindered phenol antioxidants have complex structures and lengthy IUPAC names and are more conveniently called by trade names assigned by manufacturers, *e.g.*, Irganox® 1135 from Ciba (now BASF). Structures of BHT and other hindered phenol antioxidants used for stabilization of polypropylene are shown in Figure 11.4.

The optimal quantity of an additive formulated into a resin varies with the type of polymer, the specific additive and the desired effect. For example, the quantity of anti-block additives required

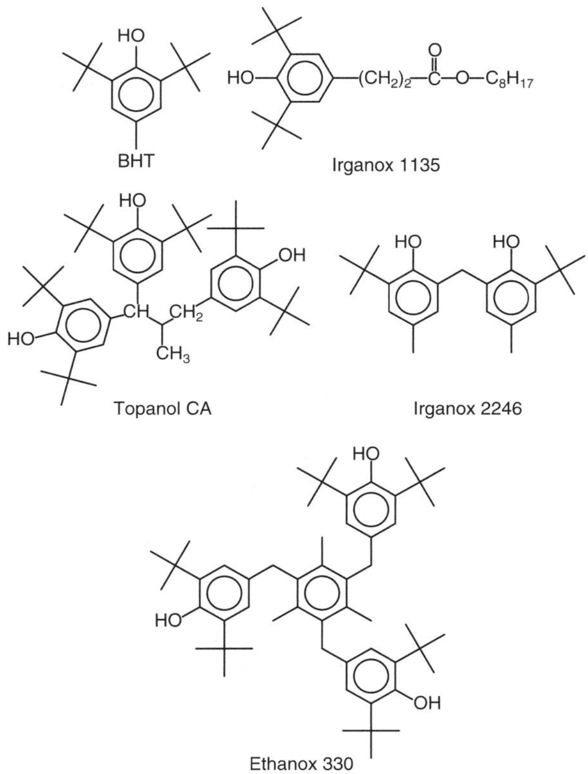

Figure 11.4 Structures of hindered phenol antioxidants. Reproduced with permission from J. Fink, *A Concise Introduction to Additives for Thermoplastic Polymers*, Scrivener Publishing, Salem, MA, 2010.

can be as high as several percent for certain polyethylene applications. ("Blocking" may be described as the tendency of sheets of polymeric film to "stick" together.) However, for polypropylene films, quantities of anti-blocking additives needed are usually in the range of 0.1 to 0.4% (1000 to 4000 ppm).

11.3   Fabrication Methods

We mentioned in Chapter 1 that fabrication methods listed below collectively account for nearly 90% of all polypropylene processed:

- *injection molding*
- *fiber extrusion*
- *film extrusion*

A more detailed breakdown of processing methods is shown in Figure 11.5.

The most suitable properties for polypropylene used in each fabrication method vary. The relationship between processing method and optimal polymer properties is complex and often involves balancing cost and ease of processing with mechanical properties needed for

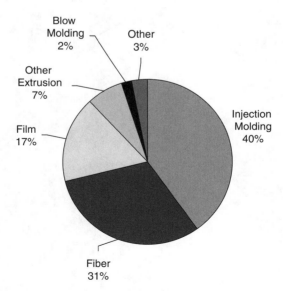

Figure 11.5 Principal processing methods for polypropylene. Source: B. B. Singh, Chemical Marketing Resources, Webster, TX, March 26, 2007.

the application. Extrusion and molding of plastics are sciences unto themselves and much has been written about these topics. In general, for injection molding applications, it is preferred to use polypropylene having relatively high melt flow rates (typically 10–20). This insures excellent flow of molten polypropylene and efficient filling of molds. Similarly for film extrusion applications, polypropylene with lower melt flow rates (higher MW) generally provides more desirable properties (See also discussion in section 2.3.4.1).

As noted earlier, virgin industrial polypropylene is especially vulnerable to oxidative degradation. However, stabilized polypropylene is resistant to oxidation and chemical attack. In most circumstances, this is regarded as an attribute. However, it also means that it is difficult to affix identifying labels or stenciling directly to polypropylene articles because glues, paints and inks do not adhere well. Several methods for surface modification have been developed to overcome this shortcoming, including gas phase treatments [12]. Treatment introduces functional groups on the polymer surface enabling bonding between polypropylene and coatings. A widely used method is called "corona discharge." This technique involves exposing the polypropylene surface to an electrical discharge that roughens the surface and generates accessible surface free radicals. These free radicals enable bonds to form between the PP surface and paints, inks and glues, resulting in improved adherence.

A technique called "controlled rheology" is also used to modify polypropylene properties. The method is sometimes called "visbreaking" (or viscosity breaking) and involves intentional use of an organic peroxide to generate free radicals and to fragment polypropylene *via* chain scission. The resultant product is called controlled rheology polypropylene (CRPP) and the overall effect is to lower the molecular weight (raise melt flow rate) and simultaneously to narrow the molecular weight distribution of the product. CRPP has improved flow properties and is more suitable for certain end-uses, particularly fiber applications. The most commonly used organic peroxide for CRPP is 2,5-dimethyl-2,5-(di-*t*-butylperoxy) hexane, skeletal structure shown in Figure 11.6. This peroxide is more commonly known by trade names such as Trigonox 101 and Luperox 101. Of course, free radical reactions are involved in both visbreaking and oxidative degradation. However there are important differences. Visbreaking occurs uniformly throughout the polymer while oxidative degradation begins primarily as an indiscriminate surface phenomenon. (For an historical view of CRPP, see

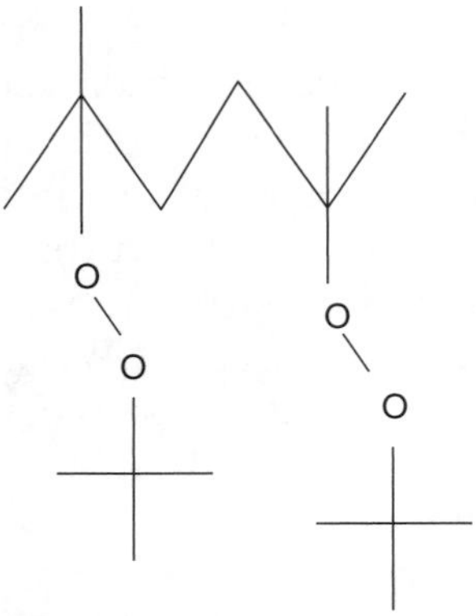

2,5-dimethyl-2,5-(di-t-butylperoxy)hexane

Figure 11.6 Skeletal structure of organic peroxide used for CRPP.

R Kowalski in *History of Polyolefins* (R. Seymour and T Cheng, editors), Reidel, 307, 1986.)

Yet another specialty method of fabricating PP film is termed "biaxially oriented polypropylene" (or "BOPP"). Film is typically stretched sequentially in orthogonal directions (termed "machine direction" (MD) and "transverse direction" (TD)) while the polymer sheet is held below the melting temperature (typically between about 120°C and 160°C). MD stretching is usually done at about 115–120°C while TD stretching is done subsequently at about 155°C. Stretching orients the polymer chains and, after cooling, results in film with improved properties, such as greater tensile strength and modulus, exceptional clarity and reduced moisture permeability. BOPP is particularly important in production of high clarity film for packaging.

11.4   Biopolymers

Thermoplastic polymers may be produced from biomass and have become known as "biopolymers." These polymers are promoted as environmental friendly alternatives to plastics derived from petroleum. Biomass may be defined simply as products obtained from biological processes, both plant and animal. In the context of modern ideological environmentalism, however, the term is often interpreted more narrowly to mean products derived from plant life. Biomass encompasses not only food crops such as corn, sugar cane, wheat, *etc.*, but, even more desirably, crop by-products such as bagasse from sugar cane and lignin from wood. Because it is believed that it will eventually not be possible to produce the requisite amounts of biomass from staple crops such as corn, soy, rapeseed, etc., algae have lately been touted as a "green" (pardon the pun) source of "biocrude" [13]. Though it is obvious that biocrude from algae will be practical only in the distant future, if biocrude does become available, it will be possible to refine using many of the same technologies employed for petroleum refining and could also provide simple building-block chemicals such as ethylene and propylene.

Biopolymers from biomass are deemed renewable and inexhaustible, though at this writing they are significantly more costly than polymers derived from petroleum. It is often suggested that improvements in technologies and economies of scale will make biopolymers economically viable in the long run. Though much more attention has been expended on the use of biomass as a source of biofuels, this segment will focus on use of biomass for making biopolymers.

Synthetic biopolymers are now commercially available and are touted as green alternatives to polypropylene and other plastics derived from coal, petroleum or natural gas. An example is poly (lactic acid). PLA is presently produced from corn, though it can also be obtained from wheat and other crops (more on PLA below). Another commercial biopolymer is "biopolyethylene" manufactured in Brazil using ethylene from dehydration of ethanol (obtained by fermentation of sugar cane) [14]. Braskem, a Brazilian petrochemical company, announced plans to produce polypropylene using propylene synthesized from ethanol [15]. Of course, Brazil has a huge capacity to produce ethanol by fermentation of sugar cane [16]. In principle, this development suggests that "biopolypropylene" from plants is also possible. However, the Braskem spokesman was evasive when asked about cost-competitiveness of the product and was quoted to have stated that the propylene *"will be derived from ethanol not directly from biomass"* (emphasis added).

Though significant technologically, it is important to keep biopolymers in perspective in today's market. Singh estimated that global production of biopolymers (including PLA) in 2007 was about 250,000 metric tons, which corresponded to only 0.25% of the global consumption of polyolefins [1]. Similarly, amounts of "biopolyethylene" and "biopolypropylene" produced from bioethanol are minuscule relative to the total amount of polyolefins manufactured.

Cost-competitiveness of green polymers will remain a substantial barrier well into the 21st century. However, green polymers will probably continue to grow despite their high cost because of political correctness and the worldwide trend of companies striving to be perceived by the public as committed to "sustainability." Also, government subsidies for green products may make biopolymers somewhat closer to cost-competitiveness in the near term. (In his book about saving the environment, Huber [17] argued against government subsidies on agricultural products, which would include "biopolymers." See also section 13.3.)

Though biodegradability is often cited as a *raison d'etre* for manufacture of green polymers, it is erroneous to conflate biopolymers with biodegradability, since not all biopolymers are biodegradable [1]. Moreover, for optimal results, many biopolymers require conditions that are conducive to biodegradation, *e.g.*, slightly elevated temperatures and contact with oxygen. (In most cases, simple exposure to ambient environmental conditions will not degrade biopolymers at an acceptable rate.)

The very definition of biodegradability and the legality of marketing claims for biopolymers are being debated. What is meant by "biodegradability" in company advertisements is often not clearly understood and can be vague. Though ASTM has standard tests for biodegradability of plastics under a variety of test conditions (*e.g.*, aerobic, anaerobic, in water), it is not yet established whether these tests can be used to certify marketing claims. Setting detailed requirements beyond pass-fail may be needed to justify a claim that a biopolymer is "legally" biodegradable. In this way, a certification could be bestowed by a third party. Such certifications may help restrain exaggerated claims of biodegradability in company marketing campaigns.

Several biopolymers are most efficiently produced from staple food crops, especially corn. However, food prices increase when crops are diverted to green products, rather than feeding the masses. Unintended consequences of using food crops to produce plastics (or fuel, for that matter) are sometimes downplayed, glibly dismissed or ignored altogether by environmental ideologues and many in the mainstream media.

For biopolymer production, biomass must first be converted to the appropriate starting materials, requiring additional processing steps. For example, ethanol must be produced by fermentation from corn or sugar cane and subsequently converted into ethylene (for biopolyethylene) or propylene (for biopolypropylene). As shown in Figure 11.7, lactic acid is first converted into the lactide to produce

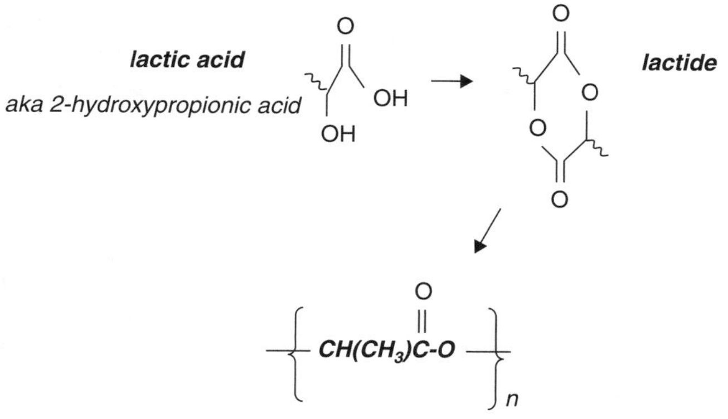

Figure 11.7 Poly (Lactic Acid) is usually produced by ring opening polymerization of the lactide obtained from corn. Note chirality of methine carbon in lactic acid. This leads to several stereochemical versions of PLA. (Reproduced from *Introduction to Industrial Polyethylene*, with permission of Scrivener Publishing LLC).

PLA. Even if these intermediate processes are considered simple and easily achieved, they nonetheless add cost.

As previously mentioned, one of the most highly developed biopolymers is PLA. In the USA, PLA is produced by NatureWorks in Nebraska from lactic acid obtained from corn. Because of the presence of a chiral carbon atom, stereochemical versions of PLA are possible. Poly (lactic acid) is produced by ring opening polymerization of the lactide, as shown in Figure 11.7. PLA is technically a polyester and its IUPAC name is rather complex. Key driving forces for PLA are that it is derived from renewable resources (its "sustainability") and its biodegradability. PLA is indeed biodegradable, but requires appropriate composting conditions to insure that degradation proceeds at an acceptable rate [18]. Biodegradation of PLA requires temperatures of about 140°F (and occasional churning to increase air-contact) for many days to insure decomposition, ultimately into CO_2 and water. Unfortunately, such conditions are not typical of landfills or backyard compost heaps.

Polyhydroxyalkanoates (PHAs) constitute another type of biopolymer that may be derived from biomass. PHAs, manufactured in the USA by Metabolix under the tradename Mirel, are commonly produced by fermentation of sugars through the action of microorganisms. Like most organic polymers, Mirel polymers do not readily biodegrade under the anaerobic conditions typical of landfills.

However, PHAs readily degrade in soil or under moist conditions in the environment. Though PHAs have attractive properties, they are presently much too costly to compete effectively with polypropylene.

Using biomass to produce fuels is still a relatively immature technology. Using biomass to produce biopolymers is even less mature. However, questions about using biomass to obtain biopolymers are beginning to be asked comparable to those historically asked about use of biomass to produce biofuels. For example, there is still disagreement as to whether transformation of biomass into biofuels is advantageous from the standpoint of energy use. Bryce has discussed at length whether producing ethanol from corn requires more energy than is obtained [19]. It remains in dispute to this date. Similarly, whether obtaining biopolymers from biomass is advantageous energy-wise is beginning to be discussed. Costs for biopolymers are today significantly higher than polymers produced from petroleum products. (At this writing, PHAs cost about $2.50/lb versus about $1/lb for polypropylene). However, it is not clear how much of the higher cost is attributable to energy consumption. The energy expended to manufacture a biopolymer versus a polymer derived from petroleum will likely be disputed for years to come. An additional point of dissension may focus on the more nebulous question concerning the wisdom of using food crops to produce biopolymers instead of foodstuffs.

11.5   Environmental

Polypropylene, like most thermoplastics, is eminently recyclable. However, at this writing, the infrastructure to collect, separate and reprocess polypropylene is very limited [20]. Economics are not now favorable for recycle and/or reuse. Consequently, most polypropylene waste in the USA goes into landfills. In Europe, a significant portion of waste plastics (including polypropylene) is incinerated to reclaim the caloric value and generate electricity. In the USA, it has been claimed that ~29 million tons of municipal solid waste (~12% of the total) were incinerated in 2009 to capture the energy content [21]. This is termed EfW ("Energy-from-Waste").

A common misconception by the public is that "plastics" are the largest contributors to landfills. However, according to EPA figures, the reality is that plastics contributed only about 12.4% of the total municipal solid waste in 2010, slightly less than the amount of food waste (see Figure 11.8). Indeed, the contribution of plastics to MSW has been nearly constant for years, hovering around 12%. Paper products have historically been the largest contributors to MSW in the USA, accounting for about 28.5% of the total municipal solid waste of 250 million tons in 2010. Figure 11.9 shows the annual percentages of paper and plastics in total MSW over the period 2005–2010, the latest years for which EPA data are available. Note

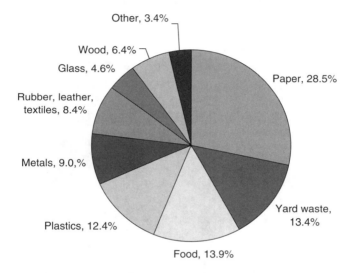

Figure 11.8 Municipal solid waste in USA in 2010. (Total MSW in 2010: 250 million tons)

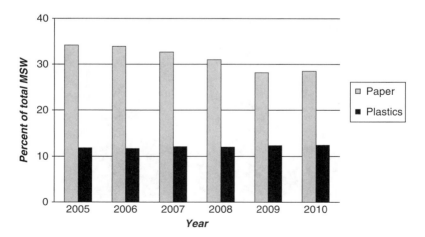

Figure 11.9 Paper and plastics in municipal solid waste.

that paper has declined slightly since 2005, while the percentages for plastics have changed little.

Since the 12.4% figure for 2010 includes all thermoplastics (LDPE, LLDPE, HDPE, PVC, PS, PETE, etc.), the actual contribution of polypropylene to total municipal solid waste in the USA is quite low, probably less than 3%. Remarkably, the overall contribution of plastics to landfills in 2010 was only marginally higher than it was in 1970 [22], owing to a variety of factors. For example, recycle rates of PETE and HDPE, though still relatively low, have become significant since 1970. Another key factor is that, in recent decades, items have been fabricated from plastics with improved mechanical strength which has permitted "thinwalling" and "downgauging" of plastic parts, resulting in lower weights per unit [22].

Figure 11.10 According to EPA data, total municipal solid waste in the USA in 2010 was 250 million tons. Of this, only ~12.4% was plastics and it is estimated that less than 3% was polypropylene.

Lack of landfill space is sometimes cited as a justification for banning plastics. (See, for example, the rationale for the San Francisco ban on plastic bags [23]). As noted above, however, plastics constituted only about 12.4% of total municipal solid waste in 2010. In fact, the overall contribution of plastics to total MSW has increased very little for ~ 40 years. Moreover, the USA is *not* running out of landfill space (contrary to comments made by a misinformed San Francisco city supervisor). Lomborg has calculated that the quantity of MSW estimated to be generated in the entire USA over *the whole of the 21ˢᵗ century* could be contained in a *single* landfill less than 18 miles square [24]. This amounts to less than 0.009% of the total landmass in the contiguous USA! Of course, a single landfill for the entire country is not a practical approach to the perceived landfill problem, but Lomborg has clearly illustrated that lack of landfill space is not a valid reason for banning plastics.

Disposal of polypropylene in landfills effectively sequesters the carbon content of the polymer. In other words, putting polypropylene in landfills is tantamount to "mummifying" the carbon content [25]. Polypropylene will remain virtually unchanged for centuries under the anaerobic conditions typical of landfills. Putting polypropylene in landfills effectively returns carbon to the soil, from whence it originated primarily in the form of oil. For those concerned about the "carbon footprint" of products and human activities, landfilling polypropylene could be viewed as the completion of the "life cycle" of carbon originally extracted from the earth. Alternatively, if polypropylene were to be incinerated, approximately three pounds of the greenhouse gas carbon dioxide would be liberated into the atmosphere for each pound of polymer combusted.

For those who accept the contention that "global warming" is being caused by carbon dioxide from burning of fossil fuels by humanity, putting polypropylene in landfills would be the preferred method of disposal, at least as compared to incineration. *[However, the principal author (DBM) of this chapter is of the opinion that global warming (or "climate change," as some prefer to call it) that may be occurring is primarily due to natural solar cycles and human-kind's contribution is minuscule. Change is precisely what climate has done for billions of years, eons before mankind even appeared. Humans have walked this Earth for a mere blink of an eye on the geologic timescale. Climate is not static; it has changed throughout the 4.5 billion-year history of Earth. As it has for time immemorial, climate will continue to get warmer (or cooler), and CO_2 levels will increase and decrease owing to* *vulcanism, and other natural causes. This will happen, regardless of man-kind's feeble, but societally highly disruptive, attempts to stop such changes through government regulations. The* cause celébre *of "saving the planet" should not require mankind to regress to the days of wood-burning stoves for home heating and horse-drawn carriages for transportation, as some ideologues would have it. While it is entirely appropriate that mankind pre-serve the environment, conserve natural resources and find new ways to fuel the future, those endeavors should not become a roundabout means of imposing a political philosophy upon society.]*

To aid recycling of plastics, the Society of the Plastics industry (SPI) has published numeric codes to identify the plastic used in fabricated articles. Each article should have an imprint of a triangle enclosing a number identifying the plastic used in its fabrication. For example, polypropylene is designated as number 5. Codes for other plastics are shown in Figure 11.10.

One of the most prominent environmental organizations is Greenpeace, known for their confrontational style over issues perceived to be important to the environment. Greenpeace has taken stances opposing certain types of chemicals, including those containing chlorine such as poly (vinyl chloride). Greenpeace has published a "Pyramid of Plastics" (Figure 11.12) that ranks plastics from least to most desirable from an environmental viewpoint. In large measure because of its chlorine content, Greenpeace places PVC at the apex of the pyramid, consistent with their assessment that PVC is the most objectionable plastic. Other problematic aspects of PVC according to Greenpeace are the plasticizers and/or

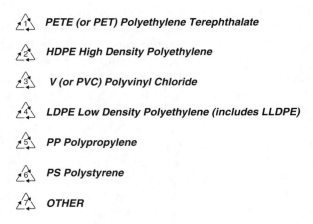

Figure 11.11 SPI coding of plastics.

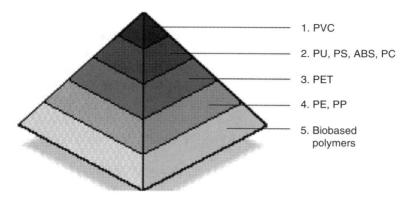

1. PVC

2. PU, PS, ABS, PC

3. PET

4. PE, PP

5. Biobased polymers

Figure 11.12 Greenpeace pyramid of plastics. Source: http://archive.greenpeace. org/toxics/pvcdatabase/bad.html Used with permission of Greenpeace.

flame retardants used. In addition, in its manufacture PVC employs vinyl chloride monomer, which is a carcinogen. Moreover, PVC burns to generate corrosive, toxic HCl. (Unlike PVC, combustion of polypropylene generates only CO_2 and water.) Greenpeace ranks polypropylene among plastics that are less objectionable (near the base of the pyramid). Of course, "biobased polymers" derived from renewable resources are most preferred by Greenpeace. Similar pyramid rankings of plastics based upon practicality or cost have not been published by Greenpeace.

Catchwords often associated with environmental sustainability are the "3 Rs" (reduce, reuse, recycle) and life-cycle analysis. Much of the discussion above relates to these aspects of polypropylene and its environmental fate. Landfilling is usually considered a last resort, though, as we have suggested, there is at least one aspect of landfilling polypropylene that even ideological environmentalists might view positively. (To reiterate, landfilling returns carbon to the soil and effectively sequesters the carbon content of polypropylene, thereby completing its life-cycle and reducing its "carbon footprint.") Life-cycle analyses are increasingly important in assessing "cradle-to-grave" impacts of products (not just for polypropylene), and the literature is replete with overviews of methodology and case studies [26].

Using landfills as a means of disposal of polypropylene (or any solid waste, for that matter) is inherently repugnant to all who wish to conserve resources. Though Lomborg has clearly shown that the US is NOT running out of landfill space, landfilling should be regarded as a last resort and municipalities should continue trying to reduce quantities of solid waste destined for landfills. Realistically, however, there will always be waste for which burial will be the only practical alternative. For example, if all municipal solid waste were to be incinerated for its heat value ("EfW"), the residual ash (including that from the ~3% polypropylene) will consist mostly of inorganic metal-containing noncombustibles and other refractory materials. These by-products of combustion will most likely need to be buried in a landfill. However, the overall quantity of solid waste for landfilling will be reduced enormously.

Society should rightly strive to minimize MSW, since that implies greater efficiency in use of resources. However, recycling is not necessarily the all-encompassing solution [27] and banning plastics is not a rational way of addressing the problem. The local SF politician who contended that plastic bags are "relics of the past" [23] failed to recognize that, contrariwise, plastics are important contributors to modern life. (It would have been more accurate had he said that *paper bags* are relics of the past; see section 13.3.) Unquestionably, research should continue to develop better, less costly and more sustainable alternatives, such as biopolymers. However, biopolymers must become more practical and cost-effective, and should succeed in the marketplace on their own merits. Their marketing should not be mandated or subsidized by government. Moreover, their introduction should cause little or no unintended consequences, thereby avoiding what occurred with food prices when vast quantities of corn were diverted to produce ethanol for use as a transportation fuel. Even if practical biopolymers are developed, it is likely that, rather than completely replacing polypropylene and other petroleum-derived plastics, biopolymers will coexist with established plastics for many years to come and fulfill a complementary role. There should be ample room in today's market place for both.

11.6   Questions

1. What is meant by the term "heat history" of polypropylene? What are the consequences of a heat history of polypropylene?
2. What are the principal reasons that additives are used for polymers?

3. What are the two most important types of antioxidants used in polypropylene?
4. Why are tertiary carbons in polypropylene more susceptible to free radical formation? Write several free radical reactions by which polypropylene may undergo degradation.
5. Give three examples of biopolymers and the renewable raw material source for its manufacture. Give two examples of biopolymers that are *not* biodegradable. What is the principal barrier to broader commercialization of biopolymers?
6. What is the contribution of thermoplastics to total municipal solid waste in the USA? What is the estimated %-age of polypropylene in MSW? What happens to polypropylene in the typical landfill?
7. Very little polypropylene is recycled, despite the fact that the polymer is easily melted and reshaped. Why?
8. Why might putting PP in landfills be a preferred way of disposing of PP by environmentalists concerned about global warming?

References

1. BB Singh, *International Conference on Polyolefins*, Society of Plastics Engineers, Houston, TX, February 22–25, 2009.
2. H. Zweifel (editor), *Plastics Additives Handbook*, Hanser Publishers, 5th edition, Munich, 2001.
3. H. Zweifel, R. Maier, and M. Schiller (editors), *Plastics Additives Handbook*, Hanser Gardner Publishers, 6th edition, Cincinnati, 2009.
4. J. Fink, *A Concise Introduction to Additives for Thermoplastic Polymers*, Scrivener Publishing, Salem, MA, 2010.
5. R.E. King, III, "Overview of Additives for Film Products", *TAPPI Polymer Laminations and Coatings Extrusion Manual*, T. Butler (editor), TAPPI Press, September, 2000.
6. P. Patel and B. Puckerin, "A Review of Additives for Plastics: Colorants" *Plastics Engineering*, Society of Plastics Engineers, November, 2006.
7. P. Patel and N. Savargaonkar, "A Review of Additives for Plastics: Slips and Antiblocks" *Plastics Engineering*, Society of Plastics Engineers, January 2007.
8. P. Patel, "A Review of Additives for Plastics: Functional Film Additives," *Plastics Engineering*, Society of Plastics Engineers, August, 2007.

9. R. Stewart, "Flame Retardants," *Plastics Engineering*, Society of Plastics Engineers, February 2009.

10. R Becker, E Burgin, L Burton and S Amos, *Polypropylene Handbook*, (N Pasquini, editor), 267 (2000).

11. C. Vasile, *Handbook of Polyolefins*, Marcel Dekker, 413 (2000).

12. M. Gheorghiu, G. Popa and M. Pascu, *Handbook of Polyolefins*, (C. Vasile, editor), Marcel Dekker, 649 (2000).

13. W Thurmond, *International Conference on Polyolefins*, Society of Plastics Engineers, Houston, TX, February 21–24, 2010.

14. AH Tullo, *Chemical and Engineering News*, p 21, September 29, 2008; see also *Chemical and Engineering News*, p. 9, July 25, 2011 and J Wooster, B Pereira and A Ulriksen, *International Conference on Polyolefins*, Society of Plastics Engineers, Houston, TX, February 22–25, 2009.

15. AH Tullo, *Chemical and Engineering News*, p. 26, November 15, 2010.

16. SK Ritter, *Chemical and Engineering News*, p. 15, June 25, 2007.

17. P Huber, *Hard Green*, Basic Books (Perseus Books), 120–122, 1999.

18. E. Royte, *Smithsonian Magazine*, August, 2006; for entire article go to www.Smithsonian.com and search for "poly (lactic acid)."

19. R Bryce, *Gusher of Lies*, PublicAffairs, 162 (2008).

20. M. McCoy, *Chemical and Engineering News*, p. 30, March 16, 2009.

21. JG Waffenschmidt, *International Conference on Polyolefins*, Society of Plastics Engineers, Houston, TX, February 27, 2011; see also M Bomgardner, *Chemical & Engineering News*, p. 14, August 1, 2011.

22. W Raftgey, *Saturday Night with Connie Chung*, AGS & R Communications, May 2, 1991.

23. C. Goodyear, *San Francisco Chronicle*, March 28, 2007.

24. B Lomborg, *The Skeptical Environmentalist*, Cambridge University Press, 207, 2001.

25. P Huber, *Hard Green*, Basic Books (Perseus Books), 115 (1999).

26. See for example M Werner, *International Conference on Polyolefins*, Society of Plastics Engineers, Houston, TX, February 21–24, 2010; R Stolmeier and D Anzini, *International Conference on Polyolefins*, Society of Plastics Engineers, Houston, TX, February 22–25, 2011; R Anderson and V Ngo, *International Conference on Polyolefins*, Society of Plastics Engineers, Houston, TX, February 22–25, 2009.

27. P Huber, *Hard Green*, Basic Books (Perseus Books), 120 (1999).

12

Overview of Polypropylene Markets

12.1 Introduction

Beginning with a few million pounds in 1957 in Italy, the USA and Germany, crystalline polypropylene has grown to become the gargantuan global business it is today. Billions of pounds are produced each year on 6 continents. Though all forms of polyethylene (HDPE, LDPE, EVA, LLDPE, etc.) if taken together remain the largest volume plastic, a recent analysis suggests that polypropylene is the *single* largest volume plastic produced globally, exceeding even that of HDPE which is the largest type of polyethylene manufactured [1] (See Figure 12.1). Polypropylene is fabricated into hundreds of consumer and industrial goods that affect virtually every aspect of daily life.

Evolution of the polypropylene business is truly an amazing story. We reviewed a few of the historical details in Chapters 1, 3 and 4 and mentioned some of the pioneers from the 1950s who paved the way for the business today. The baton has since been passed to new generations of chemists and engineers who work to deliver this versatile polymer to markets around the world. These newcomers to the polypropylene industry will include many readers of this book.

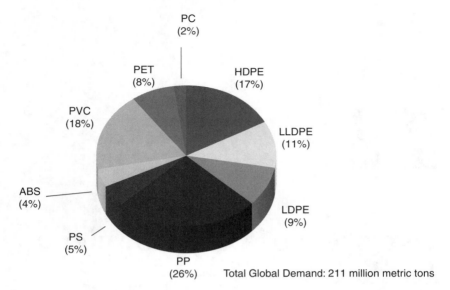

Figure 12.1 Global market shares of various thermoplastics. Source: H. Rappaport, International Conference on Polyolefins, Society of Plastics Engineers, Houston, TX, February, 2012.

In this chapter, we will survey some of the important features of the polypropylene business. We will address topics such as:

- The supply chain for polypropylene
- Global volumes of polypropylene.
- Major manufacturers of polypropylene.
- Consumer goods fabricated from polypropylene.

12.2 The Supply Chain for Polypropylene

The supply chain for the vast majority of industrial polypropylene begins with the extraction of petroleum from the earth. According to the BP statistical review, the global total production of petroleum in 2010 was about 82 million barrels *per day* (corresponding to approximately 22 billion pounds *per day*, see bp.com/statisticalreview). Crude petroleum is then refined into a wide range of products, including the key precursor (propylene, more below) for the polypropylene industry. Of course, most (>60%) of each barrel of oil is refined into transportation fuels. Combined, fuels

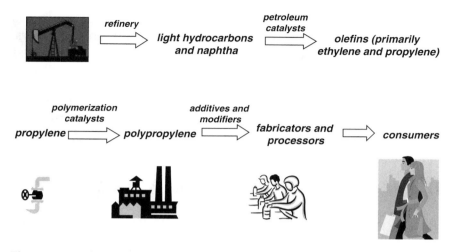

Figure 12.2 Polypropylene supply chain.

and home heating oils account for more than 80% of each barrel of oil [2]. Only a small portion (<5%) of each barrel of oil is refined into olefins (ethylene, propylene, etc.). Furthermore, ethylene and propylene have many downstream applications other than polyolefins. The schematic in Figure 12.2 illustrates steps (the "supply chain") involved in the transformation of oil into polypropylene.

The natural abundance of propylene (and olefins in general) in raw petroleum is very low [3]. This is so because of the high reactivity of olefins relative to saturated hydrocarbons, which comprise the vast majority of crude petroleum. Olefins must be synthesized from petroleum and this is most often achieved within the petrochemical industry by a technique called "cracking." (Cracking is sometimes also termed "pyrolysis" because the high molecular weight hydrocarbons undergo C-C and C-H bond cleavage caused by high temperature.) Cracking essentially converts high molecular weight hydrocarbons into smaller molecules, such as olefins.

Cracking may be accomplished by several processes. For propylene, the most important is called thermal cracking and involves free radical chemistry. Another technique for production of propylene that has gained importance in recent years is dehydrogenation of propane, a major component of liquefied petroleum gas (LPG). (LPG is probably most familiar to the public as the fuel for backyard barbecue grills.)

Petrochemistry is a vast and complex science unto itself and has evolved dramatically over the past century, especially since the 1940s. Information presented above, though greatly simplified, is intended to provide the reader a brief view of how propylene may be produced in the megaton quantities needed for the polypropylene industry. However, sources for propylene monomer may change with market conditions, availability of feed stocks, *etc.*, and can even be affected by the maneuverings of national and international politics.

12.3　The Global Polypropylene Market

In 2010, approximately 48 million metric tons of polypropylene (~106 billion pounds) were produced worldwide [4]. The names of companies that manufacture polypropylene have evolved over recent decades. When the authors of this text began their industrial careers, there were perhaps a dozen manufacturers of polypropylene in the USA alone. At that time, major producers of polypropylene were concentrated in the USA and Europe and included companies such as:

- *Amoco*
- *Arco*
- *BASF*
- *Dart Industries/Rexene/El Paso*
- *Esso (became Exxon in 1973 and ExxonMobil in 1999)*
- *Hercules (later to become Himont, Montell and later still LyondellBasell)*
- *Hoechst*
- *Montecatini (Montedison)*
- *Phillips Petroleum Co.*
- *Shell*
- *Solvay (Soltex in the USA)*

In those early days, Mitsui Petrochemical (Japan) was the dominant producer of polypropylene in Asia and was later to play a major role in the development of supported catalysts through a cooperative effort with Montedison.

A person entering the polypropylene industry within the past few years may be completely unaware of the contributions these companies made to the development of the polypropylene business. Many have since been absorbed or have morphed into modern companies such as ExxonMobil, INEOS and LyondellBasell. Consolidations and mergers significantly reduced the number of polypropylene manufacturers in the world and the dozen or so PP manufacturers competing in the USA in the 1970s have been reduced to just a handful today.

The top ten global producers of polypropylene for the years 2006 and 2010 are shown in Table 12.1 [5, 6]. The table vividly illustrates the enormous changes that the industry has undergone in recent years. It also shows the global scope of today's polypropylene business. Note that five of the top ten manufacturers in 2010 are based in the Middle East or Asia. Polymer production in the Middle East is projected to grow from about 24 million tons in 2011 to 34 million tons in 2016 [7]. In the coming years, manufacture of polypropylene is likely to continue shifting toward the Middle East and Asia.

Table 12.1 Major polypropylene producers.

Rank	Top 10 in 2006*	Top 10 in 2010**
1	Basell	LyondellBasell
2	Sinopec	Sinopec
3	INEOS	Braskem***
4	Total Petrochemical	Total Petrochemical
5	ExxonMobil	Borealis
6	SABIC	Reliance Industries
7	Borealis	CNPC
8	Reliance Industries	SABIC
9	Dow	Formosa
10	Sunoco	ExxonMobil

* RJ Bauman, Nexant ChemSystems, International Polyolefins Conference, Society of Plastics Engineers, Houston, TX, Feb 26, 2007.

** JN Swamy, Personal Communication, Chemical Marketing Resources, Webster, TX, July 29, 2011; based on published capacities.

*** As Chapter 12 was being written, Braskem (Brazil) announced the acquisition of Dow's polypropylene business (*Chemical & Engineering News*, 18, August 1, 2011).

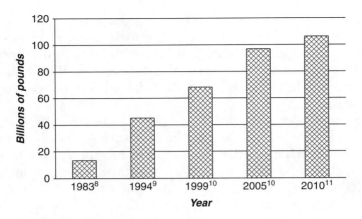

Figure 12.3 Growth of polypropylene (1983–2010).

The dramatic growth of polypropylene since 1983 is illustrated in Figure 12.3. Though data in Figure 12.3 are based on a combination of demand and capacity, the trend of strong historic growth of polypropylene is evident (see also discussion near the end of section 13.3). In the early 1970s, the growth rate for polypropylene was just over 7% per year and continued at that rate for more than 20 years [12]. The rate of growth in recent years has slowed owing to the global recession and to the maturing of some PP markets, but is expected to resume healthy growth in the coming decade.

Accompanying the changes mentioned above are dramatic shifts in the global pattern of importing and exporting polypropylene. Between 2010 and 2015, North America is projected to become a net importer and the Middle East will become an enormous exporter of polypropylene [13]. The largest portion of Middle Eastern polypropylene will go into Asia, primarily China.

Historically, growth of polypropylene has been at the expense of other materials. Products made from polypropylene have supplanted products previously made from wood, paper, glass, metals and even other polymers. That trend continues today. Reasons for displacement include lower cost and improved performance. For example, articles made from PP are often more durable than previously used materials. Because of the low density of polypropylene, articles weigh significantly less, while retaining or even improving strength. If broken, polypropylene containers do not shatter into hazardous shards, as glass does, which makes it ideal for safer packaging of many items used in the home. The relatively high

Figure 12.4 Polypropylene has many applications in household appliances including dishwashers, freezers, washing machines, etc. (Photo courtesy of Lowe's.)

melting range (>160°C) of stereoregular polypropylene means that it can be used in medical devices that require autoclaving or high temperature sterilization.

Consumer uses of polypropylene are a truly remarkable mélange of applications. Articles fabricated from polypropylene are ubiquitous in the home, at work and in transportation. Polypropylene is used to package innumerable items that appear on the shelves of your local supermarket, including microwavable containers of foods such as macaroni and cheese. It is used extensively in the automobile in the interior (*e.g.*, instrument and door panels), exterior and "under the hood" (*e.g.*, battery cases). It is found in furniture, carpeting, power tools and appliances in common use in homes and in many medical devices used in hospitals and clinics.

Figure 12.5 Polypropylene has many uses in the automotive industry, both interior and exterior. (Photo courtesy of *Plastics Engineering*, used with permission of John Wiley & Sons.)

Processing methods used for fabricating polypropylene were discussed in Chapter 11 (see section 11.3).

Uses for consumer goods may be divided into the following general categories:

- Automotive
- Appliances
- Nonwovens
- Packaging
- Consumer goods/industrial (toys, diapers, carpet backing, etc.)

Methods of fabrication vary for each category. For example, packaging applications often involve use of extruded film. Appliance and automotive parts are typically produced by injection molding. More information on each of these categories may be found on the websites of major manufacturers such as LyondellBasell, INEOS and ExxonMobil.

It is not practical to compile a comprehensive list of polypropylene applications. Nevertheless, in order to demonstrate the ubiquitous presence of polypropylene in our daily lives, Table 12.2 presents a

Table 12.2 One hundred and one polypropylene consumer products.

Artificial skating rink	Film food packaging	Photo sleeves
Audio speaker cones	Fishing line	Pitchers
Automotive battery cases	Folding saw horses	Promotional tote bags
Automotive bumpers	Food containers	Protective garden netting
Automotive door panels	Funnels	Rain suits and ponchos
Baby bottles	Golf greens turf	Rakes
Backpacks	Guitar picks	Refrigerator panels
Bandages	Hammocks	Reusable storage containers
Barrels	High chairs	Sand bags
Bicycle racks	Hinged caps	Shelving

(Continued)

Table 12.2 (cont.) One hundred and one polypropylene consumer products.

Bicycle water bottles	Home siding	Ski boots
Bins	Ice scrapers	Stackable crates
Brooms	Insulated socks	Strapping tape
Buckets	Jackets	Strollers
Chairs	Jewel cases	Student musical instruments
Child car seats	Kayaks	Surgical garments
Cigarette packaging	Labels	Swimming pool covers
Clothes hangers	Lawn edging	Swimming pools
Clothes washers	Lawn mower housing	Syringes
Combs	Long underwear	Syrup bottles
Condoms	Luggage	Tarpaulins
Coolers	Mailboxes	Tennis windscreens
Cottage cheese containers	Margarine tubs	Tool boxes
Cotton swab sticks	Mats	Toothbrushes
Currency replacement	Medical forceps	Toys
Decorative outdoor stones	Medical sutures	Trampolines
Dishwashers	Model aircraft	Utility sinks
Disposable diapers	Nautical rope	Vacuum cleaners
Disposable drinking cups	Outdoor carpets	Vehicle storage covers
Disposable flatware	Outdoor furniture	Waste baskets
Disposable medical equipment	Outdoor decking	Water filters
Dog leashes	Pet kennels	Water piping
Drinking straws	Pet mats	Yogurt containers
DVD cases		Ziplock bags

survey of 101 consumer products. The intent of Table 12.2 is to show that polypropylene plays a role in almost every facet of daily life. We use polypropylene products at mealtime, when driving a car, walking the dog, working in the office, playing sports, taking care of the baby, improving the home, maintaining our yards, brushing our teeth, visiting the doctor, playing music, and even paying bills. In so doing, polypropylene products have become integral to the quality of life in developed economies around the world.

12.4   Questions

1. What was the global total quantity of polypropylene produced in 2010? What is the projected AAGR for polypropylene for the coming decade?
2. What companies were the top ten industrial producers of polypropylene in 2010?
3. What are the major application categories for polypropylene?
4. Why is polypropylene manufacturing shifting to the Middle East and Asia?

References

1. D. Malpass, *Introduction to Industrial Polyethylene*, (published jointly by Scrivener Publishing, LLC and John Wiley & Sons), 107, 2010.
2. M. Downey, *Oil 101*, Wooden Table Press, LLC, 7, 2009.
3. K Weissermel and H.–J. Arpe, *Industrial Organic Chemistry*, (Wiley-VCH, Weinheim), 4th edition, p. 59, 2003.
4. H. Rappaport, *International Conference on Polyolefins*, Society of Plastics Engineers, Houston, TX, February, 2011.
5. C. Lee, Personal Communication, Chemical Marketing Resources, Webster, TX, November 29, 2010.
6. RJ Bauman, Nexant ChemSystems, *International Conference on Polyolefins*, Society of Plastics Engineers, Houston, TX, Feb 26, 2007.
7. J. Markarian, *Plastics Engineering*, published by the Society of Plastics Engineers, 12, June 2011.
8. JP Hogan and RL Banks, *History of Polyolefins*, (R. Seymour and T. Cheng, editors), D. Reidel Publishing Co., Dordrecht, Holland, 105, 1985; based on demand.
9. L. Tattum, *Chemical Week*, Polypropylene Review, 5, August 11, 1999; based on capacity.

10. B.B. Singh, private communication, Chemical Marketing Research, Webster, TX, March 2007; based on capacity.
11. H. Rappaport, *International Conference on Polyolefins*, Society of Plastics Engineers, Houston, TX, February, 2011; based on demand.
12. EP Moore and GA Larson, *Polypropylene Handbook*, (EP Moore, editor), Hanser/Gardner Publications, 1st edition, 257, 1996.
13. H. Rappaport, *Advances in Polyolefins 2011*, American Chemical Society, Santa Rosa, CA, September 25–28, 2011.

13

The Future of Polypropylene

13.1 Introduction

Attributes of polypropylene insure that it will remain an important part of the future of thermoplastics. Modern polypropylene is available from highly efficient processes that generate very little waste. Polypropylene exhibits highly desirable properties that make it the material of choice for an abundance of applications. Though it is lightweight, it is tough and durable. It is readily fabricated into many disparate shapes and forms. It is highly resistant to a wide range of chemicals. Its relatively high melting range allows its use in microwavable containers and in medical devices that require high temperature sterilization. Polypropylene is easy to recycle, though the infrastructure required for recycle is still in its infancy. In terms of sustainability, key factors are the availability of propylene and its chemical properties. The Plastics Scorecard, a system for assessing health and environmental aspects of plastics, rates polypropylene "A+", its maximum attainable grade, for its monomer [1].

In this chapter, we will survey key growth markets for polypropylene. The virtues of polypropylene notwithstanding, there are significant threats to the long range prospects for polypropylene. These will also be discussed in this chapter.

13.2   Key Growth Markets for Polypropylene

An important and growing use of plastics, in general, and polypropylene, in particular, is in the automotive industry [2]. In 2007, 8% of the world's polypropylene production was used for automotive applications [3]. Replacement of metal parts with lightweight, but strong, polypropylene increases fuel efficiency. This application is expected to continue strong growth.

Figure 13.1 Lightweight polypropylene from LyondelBasell is used in the rear panel of this Ford Kuga offered in the European automobile market. Applications in the transportation sector will continue strong growth because of the need to reduce vehicle weight which results in improved fuel economy. (Photo courtesy of LyondelllBasell.)

Another growing application for polypropylene is the replacement of glass in packaging. Though it varies, the density of glass is at least 2.5 times greater than that of polypropylene. Also, polypropylene is not prone to breakage, whereas glass packaging is easily broken and the resultant shards can be hazardous. Baby bottles exemplify an application where polypropylene is replacing glass.

In addition to safer packaging, substitution to low density plastics such as polypropylene reduces weight per packaging unit, lowers shipping costs, reduces energy use and lowers contribution

Figure 13.2 Polypropylene baby bottles are safe, lightweight, unbreakable, transparent, and can be molded easily into shapes ideal for a baby's small hands. Reproduced from http://www.diapers.com/p/Nuby-3-Stage-Polypropylene-Bottle-11-oz-1-pk-21177.

of packaging to municipal solid waste [4]. This trend will continue because of the drive to "sustainability."

The bright future for polypropylene is further illustrated by strong growth potential in three related applications: wood plastic composites (WPCs), natural fiber reinforced plastic (NFRP) and nanocomposites, All have small base volumes at present but are poised for strong growth in the near future. These developments bring polypropylene into the range of performance engineering plastics.

WPCs are intimate mixtures of polypropylene (or other plastics) with wood products, *e.g.*, wood flour and wood fibers. These products are formed by coextrusion of thermoplastics and wood flour with the aid of compatibilizing additives. Typical composites are about 50% wood products. They are used primarily in the building and construction industries as "plastic lumber" for outdoor decking, railings, park benches, and frames for windows and doors. Indoor applications include moldings and trim. Newer uses include such diverse applications as automotive panels, railroad ties and pencils [5, 6]. WPCs have many advantages over other building materials. For example, they are more resistant to moisture, rot and insects. Moreover, recycled polypropylene may be used for some products. The market for WPCs is growing at over 10% per year [7]. Penetration by WPCs into building materials is still quite small, <5% in 2006, so the future market potential is large [8]. By one estimate, the global market will require 3.3 billion lb of plastic by 2013 [9]. Polypropylene is about 10% of that market, corresponding to a demand of about 300 million lb/yr [10]. There are many environmental benefits:

- recycle of both scrap plastic and wood waste products,
- reduced virgin wood consumption,
- reduced energy costs in forming parts compared to solid wood,
- increased durability,
- lower maintenance and
- reduced chemicals required versus pressure-treated lumber [11].

NFRP made with virgin resin is also becoming important in building materials [12]. Historically, fiber-reinforced polypropylene has been produced using carbon and glass fibers. However, natural fibers such as jute, flax, hemp, sisal, and abaca, are expanding the range of polypropylene markets. This includes applications such as pallets where natural fibers have environmental advantages over glass fiber reinforcement, and automotive moldings where reduced weight will lower vehicle carbon dioxide emissions [13, 14].

Nanocomposites of polypropylene are materials formed by intensively blending the polymer with special small particle forms of clay or other inorganic fillers. The clay, typically montmorillinite or hectorite, is a layered structure, and the layers must be separated or exfoliated by the addition of special additives, such as quaternary ammonium compounds. The exfoliated clay is blended at several percent with polypropylene, with sizeable gains in stiffness, gas and sound barrier properties and thermal stability [15–17]. Applications being explored include automotive parts and food packaging [3, 18].

13.3   Polypropylene and Free Markets

Many of the storm clouds gathering on the polypropylene horizon stem from the possibility of government interference in free markets. For example, if development of oil and natural gas resources in the USA continues to be obstructed, cost of propylene and ultimately of polypropylene will increase in excess of the usual inflationary rates. Also, government subsidies may be used to lower the cost of biopolymer alternatives (see section 11.4) artificially, further distorting the free market.

Assuming that market forces are allowed to operate, the influence of green alternatives on the future of polypropylene is expected to be minimal for the foreseeable future. This is unlikely to change unless a breakthrough occurs that substantially lowers the cost of green polymers.

In his excellent book *Hard Green*, Huber discusses the dangers of central control on free markets and ultimately upon environmental security [19]. A concern that a "Hard Green" proponent (as described by Huber) might have is that government may mandate marketing of biopolymers. Also worrisome would be the possibility of government arbitrarily increasing costs of petroleum-derived polymers by imposition of higher taxes and/or fees in an attempt to force markets toward more costly green products. Unfortunately, those in power who wish to drive markets toward green alternatives can do so surreptitiously by delaying or outright denial of drilling and pipeline permits. Well-known recent examples in the USA are the rejection of the proposed petroleum pipeline ("Keystone") from western Canada to Texas and the prohibition of drilling in the remote and desolate Arctic National Wildlife Refuge (ANWR), deemed to be too "environmentally sensitive" for drilling. Such Keynesian interference in free market forces is a danger to the supply chain of polypropylene. Though these are largely political (rather than technological) issues, they may well dictate the future of polypropylene and other polymers derived from petroleum and natural gas.

Regrettably, in recent times negative public perception of plastics has been fostered by a cadre of misinformed journalists, politicians, and environmentalists. Though many environmentalists are sincere and well-meaning, there are also those who are ideologues and modern-day Luddites. The outcome in some communities is a pervasive and virulent anti-plastics bias, occasionally contributing to local bans on articles fabricated from plastics.

A case in point is the ban on plastic carryout bags in San Francisco [20] (mentioned in section 11.5). Though the ban was directed specifically at single-use polyethylene bags used for groceries, the episode could be considered illustrative of anti-plastics bias. Let us examine in greater detail the rationale for the SF ban and its consequences. In part, it was justified on the contention that plastic bags cause "eyesores" in the community and occupy an inordinate amount of landfill space. However, the "eyesores" of plastic bags on the roadsides of San Francisco would not exist if residents had disposed of the bags properly. Banning plastic bags because people do not properly dispose of them is not unlike banning the automobile because too many drivers exceed the speed limit. Moreover, lack of landfill space is not reality (as previously illustrated in section 11.5). As a result of the ban, consumers in SF must bring their own reusable cloth bag or use bags made from paper or biodegradable plastics.

Unfortunately for the plastics industry in general, additional local ordnances have been enacted in recent years, resulting in some forms of plastics being prohibited.

Another way local ordinances have used to steer consumers away from plastic bags is to impose a tax on each bag. However, costs of biodegradable bags can be several-fold that of PE bags. Furthermore, the alternative plastic bags may not be as "biodegradable" as advertised. (A result of the latter factoid is that, even if "biodegradable" plastic bags are used in SF, they may require weeks or months to degrade under ambient conditions and will still become "eyesores" if citizens do not dispose of them properly.)

Moreover, proponents may not have fully considered unintended consequences of such bans, including higher energy consumption and increased generation of greenhouse gases associated with producing paper bags (not to mention loss of trees) and the economic impact of the hundreds of jobs that will be lost in the industry that produces PE bags. In many cases, common-sense approaches such as reuse of plastic bags and recycling were not seriously considered as solutions to the problem. As noted above, most of the bans are directed toward "single-use" carryout plastic bags made from polyethylene film. Consequently, the impact on polypropylene for the near term will be minimal. However, such bans are symptomatic of the anti-plastics bias that permeates segments of society and could eventually affect polypropylene.

Another incipient problem for polypropylene (and indeed all technology) is the growing trend to adopt the doctrine known by the benign, but deceptive label, "The Precautionary Principle." Many ideological environmentalists strongly support The Precautionary Principle which stipulates that new technology must be rejected if it *might* harm human health or the environment. Proponents suggest that the principle be invoked "even if cause and effect relationships are not fully established scientifically" [21]. *(The authors of this text consider themselves to be "practical environmentalists", defined as those who wish to conserve the natural world while embracing technological progress. Whether a new technology is to be implemented should be determined not by fiat but by a thorough assessment of the risks and benefits to humanity and to the environment. It is rare when a new technology can be said to be completely free of risk. Had the "Precautionary Principle" been invoked in the 19th century, buffalo chips might still be used for heating and cooking, as the pioneers did in the old west!). For another viewpoint on The Precautionary Principle, see author's comments (page 571) of the 2004 novel "State of Fear" by the late Michael Crichton, a truly brilliant man taken from us too early.*

Security of supply of propylene monomer must be safeguarded to insure the future of polypropylene. As pointed out in section 12.2, the vast majority of today's propylene monomer is obtained from petroleum. However, there are several well-developed technologies available that make it possible to produce propylene from other raw materials. These include the so-called "coal-to-olefins" method, used extensively in China (which has large coal reserves), and the "methanol-to-olefins" technology, which relies upon the ready availability of methanol synthesized from natural gas. Despite uncertainties in the petroleum business (geopolitical, environmental, restricted development of domestic crude oil, etc.) supply of propylene from petroleum sources does not appear to be in jeopardy near-term.

Recent developments in the natural gas industry may have enormous beneficial impact on the long-range prospects for polypropylene. Since 2007, the natural gas industry has been revitalized by new approaches to hydraulic fracturing ("fracking") and horizontal drilling, resulting in the ability "to wring gas out of shale". [22]. These technologies resulted in a 35% increase in estimates of US reserves of natural gas between 2007 and 2009! [23]. The abundant availability of natural gas from fracking is expected to have an immediate and substantially positive effect on polyethylene. However, it is believed that it will have little or no impact on PP because most propylene comes from petroleum refining. Some have suggested that the near-term effect of the fracking revolution on propylene availability could be to make propylene in short supply and more costly. Long-term, however, the effect could be quite positive.

Though natural gas is not now widely used as a starting material for propylene, it could potentially be used in the future to produce propylene monomer by established technologies [24]. Assuming that "The Precautionary Principle" is not invoked, such technologies insure a viable supply of affordable, domestic propylene well into the 21st century. However, contamination of nearby well water with methane associated with fracking is a potential problem that will need to be resolved to safeguard the continuing utility of this technique [25]. Since other studies have found no evidence of methane contamination of groundwater caused by fracking, there is clearly a need for further research.

Another recent development that may have a significant positive impact on the future of polypropylene is the increasing use of high throughput experimental methods and robotics linked to computers. R&D on polyolefins and the catalysts employed for their manufacture will benefit from these modern innovations. These technologies make it possible to conduct hundreds of experiments per week rather than the handful that Ziegler, Natta, Hogan and other pioneers could run in the "caveman" era (early 1950s) of polypropylene research. The pace of innovation in polyolefins and catalysts will be accelerated and new catalysts, processes and products will be discovered and brought to market much more rapidly.

Yet another technological innovation that will benefit polypropylene is the emergence of a wide variety of new analytical techniques [26] such as CRYSTAF and CFC, to mention only a couple. These methods will complement the techniques discussed in Chapter 2 and greatly aid the development of new types of polypropylene and streamline process technologies for PP.

Polypropylene has historically grown at 7+% over long periods and is expected to continue growing above GDP. Recent estimates predict that the global polypropylene market will grow at 4–5% AAGR in the coming years [28]. As previously mentioned, however, manufacturing of polypropylene will continue to shift to the Middle East and Asia, especially China.

Gathering storm clouds notwithstanding, the future of polypropylene continues to be bright. Polypropylene is a versatile polymer that is easily shaped into a plethora of forms. At this point, it is available globally at reasonable cost and will continue to be a favored alternative to other materials such as metals, paper and glass. From an environmental viewpoint, Greenpeace has concluded that polyolefins are preferable to other commonly available synthetic polymers (except biopolymers) and "pose fewer risks" compared to PVC (see also discussion at the end of section 11.5). Though some may contend that polypropylene is a large contributor to the environmental problems that afflict modern life, it is more accurate to say that polypropylene will be an important part of the solution to society's environmental problems.

13.4 Questions

1. What are the advantages and environmental benefits of using wood plastic composites instead of other materials used in the construction industry?
2. Lack of landfill space is sometimes cited as a reason to ban certain plastic articles, such as San Francisco did with grocery bags. Why is lack of landfill space a fallacious reason for banning plastics?
3. What are Luddites?
4. What is "The Precautionary Principle"?
5. What are some of the reasons that Greenpeace favors polyolefins such as polypropylene over poly (vinyl chloride)?

References

1. The Plastics Scorecard is a life cycle evaluation system designed by the nonprofit Clean Production Action and Pure Strategies consulting. See: http://www.cleanproduction.org/Scorecard.Grades.php
2. M. Tolinski, *Plastics Engineering*, vol. 66, 12, January, 2010; see also ND Lamontagne, *Plastics Engineering*, vol. 66, 22, January, 2010; see also J Markarian, *Plastics Engineering*, vol. 67, 22, September, 2011.
3. S. Moritomi, T. Watanabe, S. Kanzaki, Polypropylene Compounds for Automotive Applications and reference therein. See http://www.sumitomo-chem.co.jp/english/rd/report/theses/docs/20100100_a2g.pdf
4. A. Carlson, *International Polyolefins Conference 2009*, Society of Plastics Engineers, Houston, February 22–25, 2009.
5. http://www.plasticstoday.com/articles/e-weekly-news-briefs-february-20–24 Feb. 23, 2006.
6. http://www.plasticstoday.com/mpw/articles/write-stuff-bioplastics-wpc-finds-way-writing-utensils April 12, 2010.
7. http://www.plasticstoday.com/articles/led-wpc-us-decking-demand-set-rebound-housing Feb, 8, 2011.
8. http://www.plasticstoday.com/articles/market-snapshot-building-construction-4 Sept. 30, 2007.

9. http://www.plasticstoday.com/articles/wpc-demand-growth-has-healthy-forecast Aug. 10, 2009.

10. http://www.specialtyminerals.com/specialty-applications/specialty-markets-for-minerals/plastics/wood-polymer-composites

11. http://www.plasticstoday.com/articles/would-you-believe-molded-wood, Oct. 31, 2003.

12. J Evans, *Plastics Engineering*, Vol 67, 10, July/August, 2011.

13. S.V. Joshi, L.T. Drzal, A.K. Mohanty, S. Arora, Composites, Part A, vol. 35, pp. 371–376, 2004.

14. http://www.europeanplasticsnews.com/subscriber/headlines2.html?cat = 1&id=1279700128

15. L.M.Sherman, Plastics Technology, May 1, 2007. http://www.mmsonline.com/articles/nanocomposites-less-hype-more-hard-work-on-commercial-viability

16. file http://www.intechopen.com/download/pdf/pdfs_id/14374

17. http://www.iccm-central.org/Proceedings/ICCM17proceedings/Themes/Nanocomposites/POLYMER%20NANOCOMP%20FOR%20STRUC%20APPL/E6.17%20Yan.pdf

18. M.R. Manikantan, N. Varadharaju,Packaging Technology and Science Vol. 24, pp. 191–209, 2011.

19. P Huber, *Hard Green*, Basic Books (Perseus Books), 154 (1999).

20. C. Goodyear, *San Francisco Chronicle*, March 28, 2007.

21. JH Adler, *Global Warming and Other Eco-Myths*, (Prima Publishing), R. Bailey, Ed, 265 (2002).

22. R. Bryce, *Power Hungry*, PublicAffairs™ (member of Perseus Books Group), 239 (2010).

23. JA Moore and T Shute, *The Hidden Cleantech Revolution*, Energy Publishers of America, Houston, 69 (2010).

24. G Olah, A Goeppert and G Prakash, *Beyond Oil and Gas: The Methanol Economy*, Wiley-VCH, Weinheim, 249 (2006).

25. S Ritter and G Hess, *Chemical and Engineering News*, 5, May 16, 2011.

26. N. Lamontagne, *Plastics Engineering*, Vol 67, 10, October, 2011.

27. EP Moore and GA Larson, *Polypropylene Handbook*, (EP Moore, editor), Hanser/Gardner Publications, 1st edition, 257, 1996.

28. S. Schneider, *International Conference on Polyolefins*, Society of Plastics Engineers, Houston, TX, February, 2010; H. Rappaport, *International Conference on Polyolefins*, Society of Plastics Engineers, Houston, TX, February 27, 2012.

Appendix A

Glossary of Abbreviations, Acronyms and Terminology

(Definitions of abbreviations, acronyms and terms are in context of polyolefin technology; may be different in other contexts.)

Abbreviation or Term	Definition
1,3-diethers	used as donors in new "5th generation" LyondellBasell PP catalysts
3 Rs	reduce, recycle and reuse
^{13}C	an isotope of carbon (as in carbon 13 NMR).
^{1}H	an isotope of hydrogen, a proton (as in proton NMR).
^{31}P	an isotope of phosphorus (as in phosphorus 31 NMR); used in Barron method.
α	Greek letter alpha
β	Greek letter beta
γ	Greek letter gamma
δ	Greek letter delta
η	Greek letter eta

μ	Greek letter mu; see micron
ν	Greek letter nu
π	Greek letter pi
σ	Greek letter sigma
$[\eta]$	Intrinsic Viscosity
η_{sp}	specific viscosity
AAGR	average annual growth rate (%)
ABS	acrylonitrile-butadiene-styrene terpolymer
ACGIH	American Conference of Government Industrial Hygienists (recommend TLVs for chemicals)
activator	another term for the metal alkyl cocatalyst in ZN catalyst systems
active aluminum	used to express "free TMAL" in methylaluminoxanes
activity	term for the quantity of polymer produced by a given weight of catalyst, *e.g.*, kg of PP per kg of catalyst.
aka	also known as
Al	aluminum
alkyl	generic name for hydrocarbyl groups (methyl, ethyl, isobutyl, *n*-butyl, etc.)
alpha olefins	linear 1-olefins; used as comonomers in some grades of PP copolymers.
amu	atomic mass units.
ansa metallocene	a metallocene containing a bridging ligand arrangement
anti-blocking agent	additive used to mimimize "blocking" (adhesion) of polyolefin films.
antioxidant	additive used to minimize reaction of polyolefins with atmospheric oxygen
antistatic agent	additive used to minimize static electricity in polyolefin films.
ANWR	Arctic National Wildlife Refuge
APAO	amorphous polyalphaolefins
aPP	atactic polypropylene
ARC	accelerating rate calorimetry
ASTM	American Society for Testing and Materials

atactic	amorphous form of PP; methyl groups arranged randomly along polymer chain.
atm	atmosphere (1 atm ~ 14.7 psig)
autoxidation	the reaction of molecular oxygen (from air) with PP
B	boron
Barron method	an analytical method for determining "free TMAL" in methylaluminoxanes using ^{31}P NMR.
BASF	German chemical company; name originally derived from "Badische Anilin und Soda Fabrik"
BEM	n-butylethylmagnesium
BEM-B	n-butylethylmagnesium n-butoxide
BHT	butylated hydroxy toluene, aka 2,6 di-tert-butyl-4-methylphenol
blocking	tendency of polyolefin films to stick together; problem solved by use of an anti-blocking agent
BOM	n-butyl-n-octylmagnesium
BOMAG	another abbreviation for n-butyl-n-octylmagnesium
BOPP	biaxially oriented polypropylene
bp	boiling point
BP	British Petroleum
BR	butadiene rubber
Bu	usually represents a normal butyl group ($CH_3CH_2CH_2CH_2$); see n-Bu.
BuCl or n-BuCl	n-butyl chloride (C_4H_9Cl)
bulk process	PP process wherein polymerization is conducted in liquid propylene; also called "liquid pool" process
C	Celsius (temperature in the Celsius scale)
carbalumination	addition reaction wherein a C-Al bond adds across a double bond
CASRN	Chemical Abstracts Service Registry Number
cc	cubic centimeter
CFC	cross-fractionation chromatography

CGC	controlled geometry catalyst
Cl	chloride ligand, as in Et_2AlCl.
cm^3	cubic centimeter
CMDS	cyclohexylmethyldimethoxysilane
CNPC	China National Petroleum Corporation
CO	carbon monoxide
CO_2	carbon dioxide
Co	cobalt (used in synthetic rubber and selected single-site catalyst systems).
cocatalyst	metal alkyl component of a ZN or single-site catalyst system; sometimes calld an "activator."
comonomer	an olefin other than the primary olefin in polymerizations of ethylene or propylene.
conventional MAO	MAO produced by hydrolysis of TMAL
copolymer	produced by copolymerization of 2 olefins; *e.g.*, RACO and HECO typically involve propylene and ethylene
corona discharge	gas-phase method for surface treatment of PP to improve adhesion of paints and/or inks.
Cp	cyclopentadienyl (C_5H_5) group; often a ligand in metallocene SSCs
cp	centipoise; obsolete unit of viscosity replaced by Pascals or megaPascals (mPa)
CPChem	Chevron Phillips Chemical (developed "loop slurry" process for PP)
Cr	chromium (used in silica-supported Phillips catalysts for polyethylene).
cracking	a petrochemical process in which high MW hydrocarbons are pyrolyzed to produce simpler HC, such as propylene.
CRPP	controlled rheology polypropylene; obtained by process called "visbreaking" using peroxides
CRYSTAF	crystallization analysis fractionation

CSTR	continuous stirred-tank reactor
CTA	chain transfer agents; used to control MW of polymers
D	bond dissociation energy
d	density
Dalton	unit of molecular weight, equal to the weight of a proton or neutron
DBM	dibutylmagnesium (commercial product is mixture of DNBM and DSBM).
DEAB	diethylaluminum bromide
DEAC	diethylaluminum chloride
DEAI	diethylaluminum iodide
DEAL-E	diethylaluminum ethoxide
DEZ	diethylzinc
dg	decigram (0.1 g)
DIBAC	diisobutylaluminum chloride
DIBAL-BOT	diisobutylaluminum butylated oxytoluene
DIBAL-H	diisobutylaluminum hydride
DIBAL-O	bis(diisobutylaluminum) oxide, an IBAO produced with H_2O/Al of 0.50.
DIBP	diisobutyl phthalate; used as donor in PP catalyst systems
diene	type of monomer containing two olefinic sites, e.g., 1,3-butadiene; used in prod'n of elastomers
diethers	see 1,3 diethers
dimer	state of molecular association that involves two molecules per unit.
dispersity index	see PDI
displacement reaction	see exchange reaction.
dL	deciliter
DMA	dimethylaniline (used in determining "free TMAL" content in methylaluminoxanes).
DNBM	di-n-butylmagnesium
DNOAI	di-n-octylaluminum iodide
DNPRAC	di-n-propylaluminum chloride

donor	a Lewis base, such as an ether or organic ester, used as a stereoregulator in PP catalysts.
DOT	Department of Transportation (regulates shipping containers, product classifications, etc.)
DP	degree of polymerization; number of repeating units (including end groups) in a polymer.
DPMS	diphenyldimethoxysilane; used as donor in PP catalyst systems
DSBM	di-sec-butylmagnesium
DSC	differential scanning calorimetry
EADC	ethylaluminum dichloride
EAO	ethylaluminoxane
EASC	ethylaluminum sesquichloride; $Et_3Al_2Cl_3$.
ECHR	epichlorohydrin rubber
EfW	energy from waste
EINECS	European Inventory of Existing Commercial Chemical Substances
elastomers	rubbery polymers; many made by Ziegler-Natta copolymerization of olefins and/or dienes.
ELINCS	European List of Notified Chemical Substances
EPA	Environmental Protection Agency
EPDM	ethylene propylene diene monomer; rubbery copolymer produced with ZN catalysts
EPM	ethylene-propylene monomer rubber
EPR	ethylene-propylene rubber
Et	an ethyl group; C_2H_5
EtCl	ethyl chloride; C_2H_5Cl
EtO (or OEt)	an ethoxide ligand, as in Et_2AlOEt.
EU	European Union
EVA	ethylene-vinyl acetate copolymer
exchange reaction	reaction in which an R_3Al is made by olefin exchange with a different R_3Al.
external donor	donor added separately from the catalyst in a PP catalyst system; usually added with TEAL.

extrusion plastometer	instrument used to determine MFR and FRR
FAR	film appearance rating(s); from a standard test that measures defects in polyolefin films
FDA	Federal Drug Administration
Fe	iron (used in selected single-site catalysts).
FRR	flow rate ratio
FID	free induction decay
fp	freezing point
FPO	flexible polyolefins
free TMAL	term applied to residual TMAL (or R_3Al) content of PMAOs and MMAOs
FRP	fiber-reinforced plastics
FRR	flow rate ratio; ratio of high load MFR to low load MFR. As FRR increases, MWD broadens
FTIR	Fourier transform infrared spectroscopy
FW	filtrate weight
g	gram(s)
gas phase process	polymerization process wherein particles are suspended by circulating gas
GC	gas chromatography (used for analysis of hydrolysis gas of aluminum alkyls)
GDP	gross domestic product
GPC	gel permeation chromatography; also called size exclusion chromatography (SEC)
Grignard reagent	RMgX (discovered by V. Grignard, usually in ether solution).
h	hour
hν	h (Planck's constant) times ν (frequency); expression symbolizing exposure to eletromagnetic radiation, e.g., light
halide	generic designation for bromide, chloride or iodide; often represented in molecular formulas by X.

HC	hydrocarbons, in context of PP technology usually implies aliphatic hydrocarbons such as hexane.
HCP	hexagonal close packed
HDPE	high density polyethylene (produced with ZN or Phillips catalysts).
heat history	repeated heating to melt and resolidify PP is termed its "heat history" and may be indicative of degradation
HECO	heterophasic copolymer; term commonly used for impact copolymer (aka ICP)
hemiisotactic	alternating methyl groups have an isotactic stereoconfiguration
heterogeneity index	see PDI
heterophasic copolymer	another term for impact copolymer
HIC	household and industrial chemicals
HLMI	high load MI; determined under higher weight load (21.6 kg) than MI
homopolymer	type of polymer produced with only propylene, i.e., without comonomer
HP	homopolymer
HPLC	high performance liquid chromatography (also called high pressure liquid chromatography)
HT-GPC	high temperature gel permeation chromatography
hydrogen response	term used for a catalyst's reactivity with hydrogen for MW control via chain transfer
hydride	common contaminant in R_3Al; concentration usually given as wt% AlH_3
HY-HS	high yield-high stereospecificity PP catalysts used in Mitsui Chemicals Hypol process
IBAO	isobutylaluminoxane
i-Bu	see isoBu
iC_5	isopentane
ICP	see HECO

II	isotactic index; also called total isotactic index (TII). Measure of stereoregularity of PP.
impact copolymer	PP in which relatively large amounts of comonomer are incorporated; good impact resistance
in	inch
internal donor	donor incorporated into a PP catalyst, usually supported on $MgCl_2$ (see also external donor).
iPP	isotactic polypropylene
IPRA	abbreviation for product called "isoprenylaluminum" (see ISOPRENYL).
IR	infrared
isoBu	isobutyl group; $(CH_3)_2CHCH_2$-
isoprene	common name for 2-methyl-1,3-butadiene; used in production of ISOPRENYL.
ISOPRENYL	isoprenylaluminum (same as "IPRA"); reaction product of TIBAL or DIBAL-H with isoprene.
isotactic	a stereoregular form of PP; methyl groups aligned uniformly on one side of polymer chain.
IUPAC	International Union of Pure and Applied Chemistry
IV	Intrinsic viscosity
JV	joint venture
kJ	kilojoules
kt	kilotons (1 kt = 1000 metric tons = 2.2 million lb)
L	liter
LAO	linear alpha olefins (see alpha olefins)
lb	pound
LCB	long chain branching, *e.g.*, length of alkyl side chains.
LDPE	low density polyethylene (produced with peroxides).
Lewis acid	molecular species with an empty orbital; an acceptor of electrons.

Lewis base	molecular species with available pair of electrons; a donor of electrons.
LGFPP	long-glass-fiber-reinforced-polypropylene
ligand	group or species bonded to catalyst or a metal, *e.g.*, alkyl, alkoxide, hydride, chloride, Cp, etc.
liquid pool process	see bulk process
LLDPE	linear low density polyethylene (produced with ZN, SSC or Phillips catalysts).
loop slurry process	PP process wherein polymerization is conducted in liquid propylene rapidly circulating in a pipe reactor
LPG	liquefied petroleum gas
m	meter
MA	methacrylic acid
MAGALA	from the term *magnesium aluminum alkyls*; used by Akzo Nobel for magnesium alkyls
MAO	methylaluminoxane
MD	machine direction; term used in polyolefin film testing (perpendicular to TD).
Me	a methyl group; CH_3
Met	abbreviation for metal in catalyst activity expressions
metal alkyls	products containing at least one metal-carbon σ-bond
metallocene catalysts	type of single-site catalyst derived from π-bonded organometallic compounds.
MFI	melt flow index; term sometimes misused in place of MFR (not appropriate for PP)
MFR	melt flow rate
mg	milligrams.
Mg	magnesium.
MgX_2	generic representation of a magnesium dihalide (such as magnesium chloride, $MgCl_2$)
MI	melt index; from an ASTM method; used as an indicator of MW of PE (shouldn't be used for PP).

micron (μ)	10^{-6} m
mil	10^{-3} in or ~25 microns
mileage	in context of PP, another term for the "activity" of the catalyst
min	minute
MIR	melt index ratio (HLMI/MI); an indicator of the breadth of MWD of PE; not used for PP
mL	milliliter
mm	millimeter
MMAOs	generic term for modified methylaluminoxanes; various types designated by suffix.
mmol (or mmole)	millimole (one thousandth of a mole)
M_n	number average MW
MONIBAC	(mono)isobutylaluminum dichloride; also called IBADC or IBADIC
monomer	single unit of a polymerizable molecule, e.g., ethylene or propylene.
morphology	in context of polyolefins, refers to the shape of the catalyst or polymer particle.
MPa	megapascal
mPP	polypropylene produced by a metallocene or single site catalyst
MPT	methyl p-toluate
MSDS	material safety data sheet
MSW	municipal solid waste
mt	metric ton (1 mt = 2200 lb)
MW	molecular weight
M_w	weight average MW
M_w/M_n	ratio of weight average MW to number average MW; called polydispersity index (PDI)
MWD	molecular weight distribution; key characteristic of polymers, also called polydispersity index
MZCR	multizone circulating reactor; used in LyondellBasell Spherizone technology
na	not applicable

n-Bu	a normal butyl group $CH_3(CH_2)_2CH_2-$; sometimes abbreviated Bu.
NFRP	natural fiber reinforced plastic
Ni	nickel (used in selected single-site catalysts).
NIOSH	National Institute for Occupational Safety and Health
NMR	nuclear magnetic resonance
NPL	non-pyrophoric limit (see reference 14 from chapter 5)
o	ortho (substitution pattern on benzene ring)
OAc	acetate
OPP	see BOPP
OSHA	Occupational Safety and Health Administration
organometallics	compounds that contain at least one metal-carbon bond; may be sigma or pi bond.
p	para (substitution pattern on benzene ring)
P	pressure (usually expressed in bars, atm or psig; "standard" P is 1 bar or 14.7 psig)
P5	screen size through which 5 wt% passes
P10	screen size through which 10 wt% passes
P50	screen size through which 50 wt% passes
P90	screen size through which 90 wt% passes
P95	screen size through which 95 wt% passes
PB-1	polybutene-1
PC	polycarbonate
Pd	palladium (used in selected single-site catalysts).
PDI	polydispersity index ($\overline{M_\omega}/\overline{M_n}$); also called heterogeneity index and dispersity index.
PE	polyethylene

PET (or PETE)	poly(ethylene terephthalate)
PHAs	polyhydroxyalkanoates (biopolymer made by microbial fermentation of sugars)
Phillips catalyst	silica-supported chromium catalyst for HDPE developed by Phillips Petro. in the 1950s
phthalates	esters of phthalic acid, used as 4th generation donors in PP catalysts; also used as plasticizers for PVC
PMAO	polymethylaluminoxane; a less commonly used name for MAO.
PO	polyolefins
polydispersity index	measure of MWD of a polymer; ratio of weight average MW to number average MW
polymer	a large molecule (or mixture of large molecules) consisting of repeating units of a monomer
polymerization	process whereby small molecules (monomers) are linked together to form large molecules
PP	polypropylene (produced with transition metal catalysts).
PPh_3	triphenylphosphine; used in Barron method for determining "free TMAL" in methylaluminoxanes.
ppm	parts per million
productivity	in context of PP, another term for the "activity" of the catalyst
proportionation reaction	see redistribution reaction.
PS	polystyrene, usually produced with peroxides; sPS, however, produced with SSC.
psd	particle size distribution
psig	pounds per square inch gauge
PTES	phenyltriethoxysilane; used as donor in PP catalyst systems
PUR	polyurethane
PVC	poly(vinyl chloride)

pyridine titration	name given an analytical method for determining "free TMAL" in methylaluminoxanes.
R	symbol for an alkyl group (methyl, ethyl, n-propyl, n-butyl, isobutyl, etc.)
rac	racemic
RACO	random copolymer
RCP	see RACO
R_2Mg	generic representation of a dialkyl-magnesium compound (such as DNBM)
R_3Al	generic representation of a trial-kylaluminum compound (such as TMAL, TEAL, TIBAL, etc.)
random copolymer	PP polymer in which comonomer is incorporated randomly along poly-mer chain
redistribution reaction	ligand exchange reaction which permits production of $R_nAlX_{(3-n)}$, where $1 \leq n < 3$
regioselectivity	in context of PP, direction of add'n of alkyl to propylene (primary or secondary carbon of C_3H_6?)
replication	phenomenon whereby polymer par-ticles assume the shape and psd of catalyst particles
rheology	study of the deformation and flow of fluids
RMgX	generic formula for an alkylmagne-sium halide; see Grignard reagent.
$R_nAl(OR)_{(3-n)}$	generic formula for alkylaluminum alkoxides; $1 \leq n < 3$.
$R_nAlX_{(3-n)}$	generic formula for alkylaluminum halides, dihalides, sesquihalides, etc.; $1 \leq n < 3$.
RO	symbol for an alkoxide group.
ROH	generic formula for an alcohol.
R_p	a polymeric (long chain) alkyl group
RX	generic formula for an alkyl halide, such as methyl bromide, ethyl iodide, n-butyl chloride, etc.

SABIC	Saudi Basic Industries Corporation
Salen	chelating ligand derived from the condensation of two salicylaldehydes with ethylene diamine
saponification	process in which an ester is converted to an alcohol and a carboxylic acid under basic conditions
SCB	short chain branching
SEC	size exclusion chromatography (for determining MW and MWD of polymers); see also GPC
SEM	scanning electron microscope
sesqui-	prefix signifying one and a half times, $e.g.$, EASC has an Et/Al and a Cl/Al of ~1.5
SF	San Francisco
shortstopping	term used for rapidly terminating polymerization as a safety measure, most often using CO as a catalyst poison
Si	silicon
silica	oxide of silicon (SiO_2); often used as support for PE catalysts
single-site catalysts	highly active transition metal catalysts; many (not all) based on metallocenes.
slurry process	process wherein polymerization is conducted in solvent in which polymer is insoluble and precipitates
solution process	PE process wherein polymerization is conducted in "solution" at high temperature
$Span_{90/10}$	$(P90 - P10)/P50$
$Span_{95/5}$	$(P95 - P5)/P50$
SPE	Society of Plastics Engineers
SPI	The Society of the Plastics Industry, trade association established in 1937.
sPP	syndiotactic polypropylene, made with SSC.
sPS	syndiotactic polystyrene, made with SSC.

SSC	single-site catalysts.
stereochemistry	chemistry dealing with the three-dimensional arrangement of atoms.
stereoregularity	in context of PP, has to do with orientation of methyl group in the growing polymer chain
STP	standard temperature and pressure, 273 Kelvin and 1 atmosphere
succinates	esters of succinic acid; used as donors in 5th generation LyondellBasell PP catalysts
suspension process	see slurry process
syndiotactic	PP (or PS) in which groups are oriented on alternate sides along the polymer chain.
T	temperature, in this text, exclusively in °C (Celsius scale)
t (also tert)	tertiary, e.g., a carbon atom bonded to 3 other carbon atoms
TAI	Texas Alkyls, Inc.(pioneered commercial production of aluminum alkyls via Ziegler technology in 1959)
TBAO	*tert*-butylaluminoxane
TD	transverse direction; term used in polyolefin film testing (perpendicular to MD).
TDP	total weight of dry polymer
TEAL	triethylaluminum
TEB	triethylborane
TEOS	tetraethylorthosilicate
terpolymer	copolymer in which three monomers are incoporated into the polymer
tert	see t
thermoplastic resins	polymers which can be melted repeatedly and formed into useful shapes.
thermosetting resins	polymers which, once formed, cannot be melted and reshaped
Ti	titanium (most widely used metal in ZN catalysts; also used in single-site catalysts).

TIBAL	triisobutylaluminum
$TiCl_3$	titanium trichloride ("tickle 3") prepared by reduction of $TiCl_4$; key ZN catalyst; becoming obsolete
$TiCl_4$	titanium tetrachloride ("tickle 4"); raw material for many commercial ZN catalysts
TII	total isotactic index (see also II)
TIPT	tetraisopropyl titanate
TLV	threshold limit value (recommended by ACGIH for hazardous exposure limits on chemicals)
T_m	melting point
TMAL	trimethylaluminum
TMP	2,2,6,6-tetramethylpiperidine; used as donor in PP catalyst systems
TNBAL	tri-n-butylaluminum
TNHAL	tri-n-hexylaluminum
TNOAL	tri-n-octylaluminum
TNPRAL	tri-n-propylaluminum
TP	total soluble polymer plus total weight of dry polymer (TSP + TDP)
TPO	thermoplastic polyolefin; a blend of a polyolefin (PP or PE) with a rubber such as EPDM or EPM and fillers.
TREF	temperature rising elution fractionation
TSCA	Toxic Substance Control Act (part of EPA); all chemicals are to be listed with TSCA before mfg.
TSP	total soluble polymer
TWA	time weighted average
TXI	total xylene insolubles; measure of stereoregularity of PP.
TXP	total xylene soluble polymer
Unipol	trademark for gas phase polymerization technology developed by Union Carbide (now Dow)
US (or USA)	United States of America

UV	ultraviolet
V	used by SPI as alternative abbreviation for PVC in plastic coding (PVC is # 3)
V	vanadium (used in ZN catalysts for polyethylene and synthetic rubber).
VA	vinyl acetate
vis-breaking	see CRPP
VLDPE	very low density polyethylene
$VOCl_3$	vanadium oxytrichloride ("vocal 3"); raw material for ZN catalysts
WPC	wood plastic composites
X	ligand in $R_nAlX_{(3-n)}$ or RMgX, usually signifying halide (Cl, Br or I).
XI	xylene insolubles (see also TXI)
XLPE	crosslinked polyethylene
xs	excess
XS	xylene solubles; a measure of atactic polypropylene
yield	in context of PP, another term for the "activity" of the catalyst
Ziegler-Natta catalyst	combination of a metal alkyl and a transition metalf compound; used in olefin polymerizations..
ZN catalyst	Ziegler-Natta catalyst
Zr	zirconium (widely used metal in single-site catalysts)

Appendix B

Answers to Questions

Chapter 1

1. This is a contentious question. Many (especially in Europe) insist that Giulio Natta was the discoverer of crystalline PP. Though Natta is widely credited to have been the first to prepare crystalline polypropylene, such was not the case in the US patent office. After nearly thirty years of litigation, the US courts cleared the way in 1983 for the patent office to award US Patent 4,376,851 to Hogan and Banks of Phillips Petroleum covering a product "having a substantial crystalline polypropylene content" based largely on results of an experiment Hogan and Banks conducted in June of 1951, nearly 3 years *before* Natta's preparation using a catalyst of the type discovered by Ziegler.

2. Greater than 97%. The balance is produced with single site catalysts.

3. Homopolymer (78%), random copolymer (6%) and impact (aka "heterophasic") copolymer (16%).

4. Polar comonomers such as vinyl acetate deactivate Ziegler-Natta catalysts.

5. Melt flow rate (MFR) is determined on an instrument called an extrusion plastometer. MFR is the quantity of molten polypropylene that can be extruded through a standard die, at a standard temperature (230°C for PP), with a standard load on the piston (typically 2.16 kg, sometimes called "the low load MFR," or 21.6 kg, "the high load MFR"). It is inversely proportional to MW.

6. FRR is the flow rate ratio, obtained by dividing the MFR at 21.6 kg (the high load MFR) by the MFR at 2.16 kg (the low load MFR). FRR is directly proportional to MWD, *i.e.*, as FRR increases, MWD broadens.

7. Polydispersity index (PDI) is obtained by dividing the $\overline{M}_w$ by $\overline{M}_n$. It is indicative of the broadness of MWD. As PDI increases, MWD broadens.

8. Catalyst must typically have activity greater than 150,000 kg of PP per kg of transition metal, must polymerize propylene with proper stereoregularity and regioregularity ("head-to-tail"), and must be responsive to hydrogen for MW control. If a copolymer is desired, the catalyst should also provide correct quantity and distribution of comonomer.

Chapter 2

1. $[\eta]_{vis} = KM_v^\alpha$ where K= 0.0238 mL/g, α = 0.725
   2.53 = (0.0238)*(1 dL/100 mL)*$M_v^{0.725}$ Note the units conversion for mL to dL
   Log(10630) = 0.725*$logM_v$
   5.5538 = log M_v
   M_v = 358,000

2.

Recognizing that mr=rm, the intensities are mm:mr:rr = 1:2:1.

3. rrmm

4. As the mmmm content increases, the stereoregularity of the polymer increases, which increases its chance of forming ideal crystals. Thus, melting point increases.

5. rmmm = mmmr, rrrm = mrrr, mrmm = mmrm, and rrmr = rmrr
   These represent the mirror image pairs and are accounted for in terms of the respective resonance intensities.

6.

   Every other propylene monomer is enchained in an isotactic manner.

7. For one gram of catalyst, there is 0.025 g * 0.5 = 0.0125 g catalytically active Ti
   0.0125 g * 1 mole/47.9 g = 0.000261 mole catalytically active Ti
   50,000 g PP/320,000 = 0.156 mole PP
   0.156 mole PP/(0.000261 mole Ti*2hrs)*(1hr/60 min) = 5.0 PP chains/(Ti*min)
   Thus, the average chain growth lifetime, prior to chain termination is 12 seconds.
   If the average chain MW = 320,000, then the frequency of insertion can be calculated:
   320,000 g/mole * 1 mole C_3H_6/42 g = 7619 C_3H_6 units/chain
   7619 C_3H_6 units/chain * 1 chain/12 sec = 635 C_3H_6 units/sec.
   Thus, one propylene molecule is inserted every 1/635 sec = 0.0016 sec.

8. Keeping the detectors with the column in the thermostatted cabinet shortens path length and assures that there will be no temperature gradients that could affect the polypropylene solution properties during the analysis.

9. There are three types of active centers, each with its own propagation rate and chain termination rate.

10. For uniform spheres, the size of the sphere is irrelevant.
    Density HCP = 0.7405*0.90 = 0.666 g/mL.
    Density, random packed = 0.65*0.90 = 0.585 g/mL
    One possible reason for the difference is that the PP particles are not uniform spheres. A second reason is that the polymer is not completely crystallized in the PP particle.

11. Vibrate the container as it is filled.

12. The bulk density of the whole polymer is higher than the polymer >450 microns due to the packing phenomenon shown in Figure 2.16.

13. There is only one type of active center in a metallocene catalyst, whereas there are several types of active centers in a typical Ziegler Natta catalyst.

14. Span90/10= (P90 – P10)/P50
    P90 = 891 microns
    P50 = 546 microns
    P10 = 374 microns
    Span90/10= (891 – 374)/546 = 0.958

15. The clumps and fragments may have come from catalyst particles that did not fully disperse in the polymerization medium. The heat of polymerization caused particles to fuse together, with some particles fragmenting due to a high local heating.

16. The melt flow instrument is useful. The polymer is melted and extruded through the instrument. The investigator should measure the density of the extruded polymer. Note that polymer density is not the same as bulk density.

Chapter 3

1. Ziegler had worked for many years on metal alkyls and had discovered the so-called "aufbau" reaction in which triethylaluminum reacts with ethylene to produce long chain aluminum alkyls that can be used to produce long chain α-alcohols and α-olefins. While exploring the scope of this reaction, his group serendipitously discovered that in the presence of nickel, the reaction could be stopped at the dimer stage, i.e., he could easily produce butene-1 from ethylene. This in turn impelled Ziegler to explore the influence of other transition metals on the reaction between TEAL and ethylene. This resulted in the discovery of polymerization of ethylene in the presence of titanium compounds and other transition metals. Later, Natta extended the reaction to propylene.

2. Titanium tetrachloride ($TiCl_4$) is the most widely used transition metal component for production of Ziegler-Natta catalysts because it is commercially available in great abundance at low cost and provides ZN catalysts with very high activity.

3. Prepolymerization accomplishes the following:

   a. Preserves catalyst morphology by making the catalyst particle more "robust," *i.e.*, less prone to fragmentation (which may produce undesirable fines)
   b. Insulates the catalyst particle, thereby reducing sensitivity to heat during initial stages of polymerization
   c. Increases resin bulk density.

4. The so-called bimetallic mechanism involves the aluminum alkyl in the transition state of the reaction (see eq 3.6). However, most catalyst chemists still believe the transition metal-carbon bond remains the active center for the polymerization.

5. See equations 3.7–3.10. Termination in propylene polymerization occurs predominantly by chain transfer to hydrogen (hydrogenolysis, see eq 3.7).

Chapter 4

1. For each lb catalyst 50,000 lb PP will be produced.
   Total propylene $= 50,000$ lb $\div 0.40 = 125,000$ lb propylene
   125,000 lb propylene $\times 1 \times 10^{-6}$ lb H_2O/lb propylene $= 0.125$ lb H_2O to be scavenged
   0.125 lb $H_2O \div 18$ lb H_2O/lb-mol $= 0.0069$ lb mol H_2O
   1.0 lb catalyst $\times 0.025$ lb Ti/lb catalyst $= 0.025$ lb Ti
   0.025 lb Ti $\div 47.9$ lb/lb-mol $= 0.00052$ lb-mol Ti
   TEAL $= 100$TEAL/Ti $\times 0.00052$ lb-mol Ti $= 0.052$ lb-mol TEAL
   By eq. (14) two moles of TEAL will be consumed per mole H_2O.
   Thus, fraction of TEAL consumed $= (0.0069 \times 2)/ 0.052 = 0.27$
   So, one can conclude that very low moisture impurities in propylene are required in order to preserve the TEAL cocatalyst.

2. Iron catalysts can result in a polymer that is colored because iron oxide has a brown color. Aniline is highly toxic, mutagenic, and, because it is odoriferous, it may give a foul smelling polymer.

3. Let X = lb A polymer that gives 35% bulk volume.
   Let V_A = volume occupied by A polymer.
   It takes catalyst A four hours of production and one hour of turnaround time, so throughput of A = X/ 5 hrs = 0.20X
   Catalyst S has 3000/2000 = 1.5 time the activity of A
   Therefore, the time for S to produce X lb = 4 hrs × 2000/3000 = 2.67 hrs.
   However, the bulk density of polymer from S = 0.50 g/mL,
   So, V_S = 0.40/0.50 * V_A = 0.80 V_A
   Therefore polymerization can continue, until V_A is reached:
   2.67 hrs × 1/0.80 = 3.34 hrs
   And polymer = (0.50/0.40) X = 1.25 X
   Throughput of S = 1.25X/(3.34 hrs + 1.0 hr) = 0.29X/hr.
   Thus, the throughput of S = 0.29/0.20 = 1.45 throughput of A

4. For safety reasons the dilute solutions still need to be handled as if they are pyrophoric because of the possibility of ignition, even if it is substantially reduced. For quality reasons the solutions still need to be handled under anaerobic conditions. The dilute solutions also add purchasing cost and add inert hydrocarbon into the polymerization system that will need to be disposed in some manner.

5. Hydride content in TIBAL is an order of magnitude higher than in TEAL. One may expect that more external donor will be required because of the reaction of the organosilane with hydride. Also, on a cold day TIBAL may freeze, creating production and safety issues.

6. The sphere is the geometric shape with the lowest surface area per unit volume. Stickiness requires from interparticle contact. Therefore, assuming sticky copolymer is present uniformly throughout the sphere, for a given level of sticky copolymer the fraction present at the surface will be at a minimum compared to other particle shapes.

   Further, the surface area, as a fraction of the total volume, can be expressed in terms of the radius, r, of the sphere:
   surface area of a sphere = $A = 4\pi r^2$
   volume of a sphere = $V = 4/3\pi r^3$
   $A/V = 4\pi r^2/4/3\pi r^3 = 3/r$

   Therefore, the surface area, as a fraction of the volume, decreases by the inverse of r, and larger spheres are more efficient than smaller spheres at sequestering the sticky polymer phase away from the surface.

7. 20,000 cal/mole ÷ 42 g C_3H_6/mole = 476 cal/g
476 cal/g ÷ 0.4 cal/g /°C = 1190°C !

Effective temperature control is critical for high activity catalysts where local temperature can rise quickly due to the heat from rapid polymerization.

Chapter 5

1. Organometallics are compounds that contain a direct carbon-metal bond. They encompass σ-bonded metal alkyls and π – bonded metallocenes. Metal alkyls, especially aluminum alkyls, are essential to the functioning of Ziegler-Natta catalysts and metallocenes are widely employed in single site catalyst systems.

2. Using 100 g of TEAL for purposes of calculation, the composition would then be 94.1 g (or 0.824 mole) of TEAL, 5.3 g (or 0.026 mole) of TNBAL and 0.6 g (or 0.020 mole) of AlH_3. Assume that redistribution of hydride with TEAL occurs *via* the equation below:

$$AlH_3 + 2\ Et_3Al \rightarrow 3\ Et_2AlH$$

This means that each mole of AlH_3 converts 2 moles of TEAL into 3 moles of Et_2AlH. The adjusted composition would then be 0.784 mole of TEAL, 0.026 mole of TNBAL and 0.060 mole of Et_2AlH. The corresponding weight % composition (rounded to nearest tenth) would then be 89.7% TEAL, 5.2% TNBAL and 5.2% Et_2AlH.

3. Aluminum alkyls function in several ways. The most important are:

   a. To reduce the transition metal to a lower oxidation state.
   b. To alkylate the reduced transition metal to form the active center for polymerization.
   c. To scavenge catalyst poisons (such as CO_2, water and oxygen).

   See equations (5.11)–(5.13) and discussion in section 5.6.

4. DEAC does not perform as well as TEAL with most modern supported ZN catalysts. TIBAL, while it performs as well as TEAL

in many ZN catalyst systems, contains much less aluminum than TEAL and historically has been more costly than TEAL *on a contained aluminum basis.*

5. Most organometallic contaminants in TEAL do not cause problems in ZN catalyst systems. An exception is alkylaluminum hydride species (expressed analytically as AlH_3) which can lower activity and stereoregularity with certain PP catalysts that use alkoxysilanes as external donors. (See discussion near the end of section 4.10). Other R'_3Al (where $R' \neq$ ethyl) in TEAL have no perceptible effect on catalyst activity or polymer isotactic index at least up to about 15%.

6. PP processes typically operate below 100°C, where TEAL, the most commonly used aluminum alkyl cocatalyst, is thermally stable.

7. Calculations show that TEAL with the reported analysis (95.0% TEAL, 5.0% TNBAL and negligible hydride) has an expected aluminum content of 23.1%. The reported value is 22.5%, indicating that the product contains a diluent which has lowered the aluminum content by about 3.6%.

Chapter 6

1. The atactic polypropylene is low melting and soluble in hydrocarbons, so it would normally be produced as a melt or solution. Thus, the particle size control provided by a support is irrelevant.

2. Unsupported MAO may survive, coordinate with subsequently added metallocene, and form catalyst particles that produce polymer dust and/or coat the polymerization reactor walls.

3. The donors may compete to occupy coordination sites on the SSC, reducing its overall activity.

4. Assuming trimethylaluminum is used, the expense of separate preparation of MAO is saved. Instead, it is prepared in-situ on the catalyst support. If other aluminum alkyls are used to produce the corresponding aluminoxanes, then the cost savings are even greater. However, that assumes equal polymerization performance, which is typically not the case.

5. The ion pair may be extracted from the surface of the support, causing formation of an undesired, dusty polymer morphology.

The covalent bond would remove an available coordination site for the growing polymer chain.

6. The chiral structure on the right produces isotactic polypropylene, whereas the structure on the left has two mirror planes and produces atactic polymer.

7. The bridging atom is a chiral center.

8. Metering the solid hydrated salt into TMAL solution is more difficult than metering a liquid. Incomplete reaction of the metal hydrate can also be a problem in a multiply hydrated molecule. Another potential problem is the agglomeration or clumping of solids in the reactor which can result in encasement of unreacted hydrate that will react violently and uncontrollably if exposed to TMAL. There is also a need to separate a relatively large amount of the metal salt from the final MAO solution. Finally, contamination of the MAO with small amounts of the metal might lead to MAO performance issues.

Chapter 7

1. The aromatic ester is employed in the titanation step anyway. By contrast, polar compounds are removed during the washing steps. So, the ester is much more of a concern for the recovered hydrocarbon. One possible solution is to use this hydrocarbon only for the first washing step, when the highest concentration of polar impurities are being removed, and then use clean hydrocarbon for subsequent washing steps.

2. $200/1 lb catalyst/40,000 lb PP = $0.0050/lb PP. Since it has been assumed that commodity grades of PP sell for ~$1/lb, the catalyst cost is 0.5% of the PP selling price. This is exclusive of the aluminum alkyl cocatalyst and external donor costs. From this it is clear that the catalyst can be an expensive specialty product without seriously impacting polymer selling price.

3. 500 g PP/(50000 g PP/g cat) = 0.01 g cat = 10 mg cat

   The implication is that the sampling protocol must begin with a bottle of catalyst which is representative of the batch and completely homogeneous. Representative catalyst sampling has to take into account any variations during the course of drying and any tendency toward particle size segregation in the sampling container. Even static electricity can become a factor for sampling at the 10 mg scale. The reliability of sampling is improved by duplicate polymerization testing.

4. The purity of the recovered solvent and the effect of impurities upon the catalyst will dictate the decision. For example, separate storage is indicated if the impurities cause a moderate deleterious effect so that recovered solvent may only be used as a certain percentage mixed with virgin solvent. Alternatively, it may be suitable for an early step in the process, but not a later step. If the impurities are compatible with one catalyst grade, but not another grade, separate storage is also advisable.

5. From a safety perspective the top delivery is safer. In the event of any line rupture or pump failure, the main contents of the tank will be protected from spilling. With the bottom gravity delivery a line rupture risks emptying the entire contents of the tank into the environment.

6. The first place to check is the quality of the titanium tetrachloride.

7. The reader may have thought of a number of options. Here are three to consider: 1) operating the column at a lower pressure would reduce the necessary temperature, which might reduce the tendency for tar formation. 2) leaving more recoverable $TiCl_4$ in the residue discharge might reduce its viscosity, which would reduce plugging. 3) better insulating the discharge line may reduce cooling of the heavy residues, which also reduces viscosity.

Chapter 8

1. Early polypropylene manufacturing plants used 1st generation catalysts and DEAC cocatalyst in slurry ("suspension") processes.

2. Because early plants used low-activity 1st generation catalysts, the resultant polymer had to be treated to remove catalyst residues. Also, since these antiquated catalysts showed low stereocontrol, it meant that the resin had to be extracted to remove atactic PP. Early processes required additional equipment to accomplish these steps. Modern plants use 4th and 5th generation catalysts and the polymer does not require post-reactor treatment.

3. As supported catalysts began to emerge in the 1970s, TEAL typically performed better than DEAC in PP processes. TEAL also had the advantage of being chloride-free. Use of TEAL eliminated the problem of corrosion of polymer-processing equipment caused by chloride-containing residues from the aluminum alkyl cocatalyst.

4. First, calculate the quantity of titanium needed for one year of operation:
(550,000,000 lb PP/y)(1 lb cat/40,000 lb PP)(0.031 lb Ti/1 lb cat) (1 lb-mole Ti/47.9 lb Ti) = 8.90 lb-mole Ti/y
Since Al/Ti = 80, the amount of TEAL required would be 80 × 8.90 lb-mole or 712 lb-mole TEAL which amounts to about 81,300 lbs per year.

5. The residual amount of titanium in the polymer in problem 4 will be (8.90 lb-mole Ti)(47.9 lb Ti/lb-mole Ti) = 426 lb Ti in 5.5×10^8 lb PP or about 0.8 ppm Ti
Similarly, for residual aluminum content:
(712 lb-mole Al)(27.0 lb/lb-mole Al) = 19200 lb Al in 5.5×10^8 lb PP or about 35 ppm Al

6. CO blocks coordination sites on active centers thereby stopping polymerization.

Chapter 9

1. A $TiCl_3$ catalyst is more sensitive to oxygen because the titanium is in the +3 oxidation state and can be oxidized to +4 oxidation state.

2. The rotation of the stir bar, in contact with the bottom of the reactor, gradually ground catalyst particles into small fragments.

3.

Material	Amount	Units	$/Unit	Material Cost, $
Magnesium ethoxide	15	lb	5.00	75
Titanium tetrachloride	50	liters	1.00	50
Di-n-butyl phthalate	2.00	lb	1.00	2
Toluene	840	liters	1.00	840

(Continued)

Material	Amount	Units	$/Unit	Material Cost, $
Heptane	520	liters	1.00	520
Catalyst	15	lb	99	1487
Total Solvent*/Catalyst	94	liters/lb		

*solvent = titanium tetrachloride + toluene + heptane

There is about $100 in raw materials cost per pound of catalyst, and over 90 liters in solvent use per pound of catalyst. Assuming that 1 liter is ~ 2 lb, there would be around 200 lb in waste for every pound of catalyst, without any recycling.

4. Cascade the washes; use the second wash solvent from the first batch, as the first wash solvent for the second batch, and so on.

5. The storage tank of spent liquids from the washes must be kept warm, or solids will precipitate on the interior walls.

6. The electric mantle is more prone to hot spots than the oil bath, and when the process is scaled up, the heating system will more likely resemble a hot oil bath than an electric heating mantle.

7. An oil bath is preferred for reasons of safety. If the glass reactor is accidentally broken $TiCl_4$ and DEAC will not react with the oil bath, but each would react instantly on contact with an aqueous cooling bath.

Chapter 10

1. $0.025\,kg\,Ti/kg\,catalyst \div 0.0479\,kg/mole = 0.52\,mole\,Ti/kg\,catalyst$
   At 100% active Ti, 0.52 mole CO × 30 g/mole × 0.5 = 8 g CO required to kill half the catalyst
   $8\,g\,CO/40000\,g\,PP \times 10^6 = 200\,ppm\,CO$
   At 10% active Ti, 20 ppm CO required to kill half the catalyst
   These values are much higher than the maximum CO in Table 10.1. This reflects the requirement that the impurities must have a negligible effect upon the polymerization performance, and that only a small fraction of the titanium is catalytically active.

2. 0.25 kg Ti/kg catalyst ÷ 0.0479 kg/mole = 5.2 mole Ti/kg catalyst
   At 100% active Ti, 5.2 mole CO × 30 g/mole × 0.5 = 80 g CO
   required to kill half the catalyst
   80 g CO/5000 g PP × 10^6 = 16000 ppm CO
   The importance of propylene purity is less critical with $TiCl_3$
   catalysts than supported catalysts because there is more titanium and catalyst productivity is lower.

3. Perform a slurry polymerization test using a mass flow meter
   on the propylene feed line and keeping the polymerization
   pressure constant. Vary the polymerization temperature in
   increments, monitoring the effect upon the propylene uptake
   rate.

4. Aluminum alkyls react rapidly with alcohols to form aluminum
   alkoxides. However CO is not reactive with aluminum alkyls.

5. Let V_{H2} = volume of H_2 consumed at STP
   0.02 g cat × 10,000 g PP/g cat = 200 g PP
   200 g PP × 1 mole/100,000 g = 0.002 mole PP
   0.002 mole PP requires 0.002 mole H_2
   V_{H2} = nRT/P, n = 0.0821 liter-atm/°K-mol, T = 273°K P = 1 atm.
   V_{H2} = 0.045 liters = 45 mL
   This is ~ 1/10 of the H_2 charged. Thus, H_2 is essentially constant over the polymerization.

6. Keeping the final pressure in the polymerization at 120 psig, the
   pressure due to propylene is increased by ~ 15 psi. Thus, the
   polymerization rate should increase.

7. Add xylene to the reactor and heat to reflux for several hours.
   Discharge the xylene while still hot, being careful to maintain
   proper ventilation during discharge.

8. The cocatalyst acts as a poison scavenger to purify the hexane
   prior to introduction of the catalyst. This helps preserve catalyst
   activity.

9. The external donor is used in Ziegler-Natta catalysts to deactivate atactic sites present on the catalyst surface. In metallocene
   catalysts all the catalyst sites are, in principle, the same, and
   designed to produce the desired polymer tacticity. Thus, there
   are no undesired catalyst centers to deactivate.

Chapter 11

1. Manufacturers of PP must melt the resin so it can be converted into pellets for shipment. Processors also melt the polymer to introduce additives. The cycle of repeated melting and cooling of a polymer is called its heat history. Exposure of a polymer to high temperatures can cause bond cleavage. It typically lowers MW and can degrade mechanical strength of the polymer.

2. Additives are introduced:

   a.   to stabilize the polymer,
   b.   to make the polymer easier to process, and/or
   c.   to enhance its end use properties.

3. The two most common antioxidants for polypropylene are hindered phenols and organophosphites.

4. The bond energy between a tertiary carbon and hydrogen is weaker than a secondary or primary C-H bond. Homolytic scission can therefore proceed more readily with a 3° carbon in polypropylene. See reactions in Figure 11.1 for equations representative of polymer degradation.

5. Biopolymers and their "renewable" sources include the following:

   a.   poly (lactic acid) from corn
   b.   poly hydroxyalkanoates from fermentation of sugars
   c.   polyethylene made from ethylene obtained from sugar cane
   d.   polypropylene made from propylene obtained from sugar cane

   Items c and d will not be biodegradable.

   Cost is the most serious barrier for biopolymers. Presently, biopolymers are substantially more expensive than polypropylene made from petroleum-derived raw materials. Unless a breakthrough occurs that dramatically lowers the cost of biopolymer production, this is unlikely to change in the near term.

6. Plastics constituted about 12.4% of the total municipal solid waste in the USA in 2010. It is estimated that polypropylene contributed less than 3% of the total MSW in the USA in 2010.

Essentially nothing happens to polypropylene under the anaerobic conditions of a landfill. Polypropylene will not appreciably degrade for centuries. However, landfilling PP does lower its "carbon footprint" and completes its life cycle, since it effectively sequesters the carbon content of polypropylene in the Earth, from whence it was extracted in the form of petroleum.

7. Though articles made from polypropylene are easily melted and the PP reshaped into a variety of other forms, the infrastructure to gather, separate and reprocess the polymer is not in place. Also, complexities introduced by the presence of additives (antioxidants, fillers, colorants, etc.) can make it difficult to use the recycled PP. There are presently few marketable products that can be fabricated from recycled PP in a cost-effective manner.

8. Disposal of polypropylene in landfills effectively sequesters the carbon content of the polymer. Polypropylene will remain virtually unchanged for centuries under the anaerobic conditions typical of landfills. This essentially completes the life cycle of carbon originally extracted from the earth and lowers the so-called "carbon footprint" of polypropylene. However, if polypropylene were to be incinerated, approximately three pounds of the greenhouse gas carbon dioxide would be liberated into the atmosphere for each pound of polymer combusted.

Chapter 12

1. Estimates vary somewhat, but a reasonable estimate is about 48 million metric tons (or about 106 billion pounds) in 2010.

   A more recent estimate suggests that approximately 55 million metric tons (~121 billion pounds) of polypropylene were manufactured globally in 2011.

   Estimates place the AAGR for polypropylene at 4-5% for the coming decade.

2. See Table 12.1 for major producers of PP in 2010.

3. Major categories for PP applications are:

   a. Automotive
   b. Appliances
   c. Nonwovens

d. Packaging
e. Consumer goods/industrial (toys, diapers, carpet backing, etc.)

4. Plants in the Middle East have a raw material advantage as the cost of propylene monomer is lower relative to North America and Europe, the historic locations for PP manufacture. Because of the emerging economies of China and India, Asia promises to become an enormous market for PP.

Chapter 13

1. WPCs have many advantages over other building materials. For example, they are more resistant to moisture, rot and insects. Environmental benefits of using WPCs include the following:
   • recycle of both scrap plastic and wood waste products,
   • reduced virgin wood consumption,
   • reduced energy costs in forming parts compared to solid wood,
   • increased durability,
   • lower maintenance and
   • reduced chemicals required versus pressure-treated lumber.

2. Of the 250 million tons of municipal solid waste in the USA in 2010, only 12.4% was plastics and polypropylene's contribution to MSW was estimated to be less than 3%. Moreover, the space needed for landfilling ALL of the MSW to be generated in the USA over the next century would fit into a single landfill requiring only 0.009% of the land in the contiguous US.

3. The term "Luddites" was applied to British workmen in the early 19th century who rioted and destroyed textile machinery because they believed that such machines would replace workers. The term "Luddites" has come to mean persons who oppose technological progress.

4. There are many expressions that attempt to define "The Precautionary Principle." One succinct statement is as follows: "precautionary measures should be taken even if some cause and effect relationships are not fully established scientifically." "Precautionary measures" include regulations or outright banning of certain products or technologies (*e.g.,* genetically

engineered crops). Proponents suggest that the principle should be applied in assessing environmental risks of all new technologies. It is a policy that environmental ideologues are campaigning to have included in national and international agreements. They have succeeded in a few instances, especially in Europe. However, there are many who believe that the principle is anti-science and that implementation of it will greatly hinder technological progress.

5. Greenpeace objects to PVC for a variety of reasons including:

    a. PVC is manufactured from the monomer vinyl chloride, a gaseous organochlorine compound that is a known carcinogen. The ACGIH threshold limit value (TLV) for vinyl chloride has been set at 1 ppm (8-h TWA).

    b. PVC employs additives such as phthalates and metallic compounds which are deemed potentially harmful to human health.

    c. Incineration of PVC results in liberation of toxic, corrosive hydrogen chloride and other toxic chlorine-containing compounds.

In contrast, polypropylene production employs propylene rather than a highly toxic, volatile organochlorine compound. The TLV for propylene is 500 ppm. PP is produced with a catalyst typically manufactured from $TiCl_4$, an inorganic chlorine-containing liquid, and an aluminum alkyl. This reaction affords a nonvolatile solid which becomes the active catalyst. Though $TiCl_4$ is toxic, a TLV has not yet been set by ACGIH or OSHA. (However, because it generates HCl in moist air, a key manufacturer suggests that the ACGIH "ceiling value" for HCl (2 ppm) be used). Since $TiCl_4$ is used in very small amounts, exposure is much less likely relative to the possibility of exposure to vinyl chloride during PVC production. Modern industrial PP catalysts have very high activities (often >60000 pounds of polymer per pound of catalyst). Consequently, residual chlorine levels in the final polymer are very low (ppm levels compared to approximately 57% in PVC). Further, polypropylene does not employ phthalates as plasticizers (as is the case for PVC). Moreover, incineration of PP results only in carbon dioxide and water.

Appendix C

Registered Trademarks

Akzo Nobel

Trigonox
MAGALA

Arkema

Luperox

BASF

Lynx

Braskem

INSPIRE

Becton Dickinson

LuerLok

Borealis

Borstar

Chemtura

BOMAG

ChevronPhillips

Marlex

Dow

Unipol
Versify

DuPont

    Nomex
    Teflon

Evonik

    Catylen

ExxonMobil

    Achieve
    Vistamaxx

Haver Tyler

    Ro-Tap

INEOS

    Innovene

LyondellBasell

    Catalloy
    Clyrell
    Pro-Fax
    Spheripol
    Spherizone

Mitsui

    TAFMER

Sigma-Aldrich

    Sure/Seal

Solvay

    Fortilene

Unilever

    Vaseline

Index

Also of Interest

Introduction to Industrial Polyethylene
Properties, Catalysts, and Processes
By Dennis Malpass
2010, ISBN 978-0-470-62598-9

"I found this to be a straightforward, easy-to-read, and useful introductory text on polyethylene, which will be helpful for chemists, engineers, and students who need to learn more about this complex topic."
R.E. King III; Ciba Corporation (part of the BASF group)

This concise primer reviews the history of polyethylene and introduces basic features and nomenclatures for this versatile polymer. Catalysts and cocatalysts crucial to the production of polyethylene are discussed in the next few chapter. Latter chapters provide an introduction to the processes used to manufacture polyethylene and discuss matters related to downstream applications of polyethylene such as rheology, additives, environmental issues, etc.

Introduction to Industrial Polyethylene Provides:
- A concise overview teaching the fundamentals of what polyethylene is, how it's made and processed and what happens to it after its useful life is over
- Identifies the fundamental types of polyethylene and how they differ
- Lists markets, key fabrication methods, and the major producers of polyethylene
- Provides biodegradable alternatives to polyethylene
- Describes the processes used in the manufacture of polyethylene
- Includes a thorough glossary, providing definitions of acronyms and abbreviations and also defines terms commonly used in discussions of production and properties of polyethylene.

Antioxidant Polymers
Synthesis, Properties, and Applications
Edited by Giuseppe Cirillo and Francesca Iemma
2012, ISBN 978-1-118-20854-0

Antioxidant Polymers provides a complete and detailed overview of the recent development in the field of polymeric materials showing antioxidant properties. Research into antioxidant polymers has grown enormously in the last decade because they have demonstrated a wide range of applications, from materials science to biomedical, pharmaceutical, cosmetic and personal care, as well as the food packaging industry.

After an introductory overview on the antioxidant compounds, the volume goes into the description of the natural and synthetic polymeric antioxidants in detail, with a particular attention to both their chemical and biological properties. The extraction and modification of naturally occurring polymers, as well as the fabrication of totally synthetic compounds, is treated as well. The 15 chapters are all written by acknowledged experts from both the university research environment and industry research and development labs located around the world.

The volume will be of prime interest to a wide variety of scientists and engineers including those in biomedical, pharmaceutical and food research groups; polymer science and materials science research groups; research and development divisions in the pharmaceutical, cosmetic, plastics materials, and food industries; botanical and marine researchers as well as nano engineers.